W9-BRI-177

THE RULES OF WORK

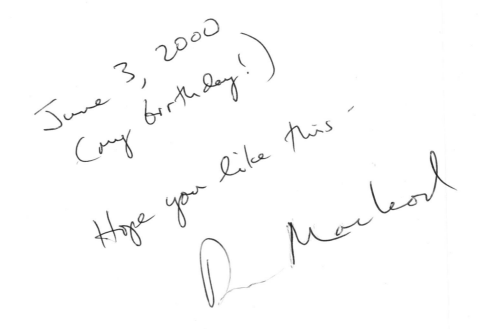

June 3, 2000
(my birthday!)

Hope you like this—

D. Macleod

THE RULES OF WORK

A Practical Engineering Guide
to Ergonomics

by Dan MacLeod

Taylor & Francis
New York

USA Publishing Office Taylor & Francis
29 West 35th Street
New York, NY 10001-2299
Tel: (212) 216-7800

Distribution Center Taylor & Francis
47 Runway Road, Suite G
Levittown, PA 19057-4700
TEL: (215) 269-0400
FAX: (215) 269-0363

UK Taylor & Francis
11 New Fetter Lane
London EC4P 4EE
Tel: 011 44 207 583 9855
Fax: 011 44 207 842 2298

The Rules of Work: A Practical Engineering Guide to Ergonomics

Copyright © 2000 by Taylor & Francis. All rights reserved. Except as permitted under the United States Copyright Act of 1976, no part of this publication may be reproduced or distributed in any form or by any means, or stored in a database or retrieval system, without prior written permission of the publisher.

1 2 3 4 5 6 7 8 9 0

A CIP catalog record for this book is available from the British Library.

Library of Congress Cataloging-in-Publication Data

MacLeod, Dan.
The rules of work: a practical engineering guide to ergonomics / by Dan MacLeod.
     p.   cm.
     Includes bibliographical references(p.) and index.
     ISBN 1–560–32885–1
     1.   Human engineering.
     I. Title.

TA166 .M28 1999
620.8′2–dc21

                                                                        99-043754

Printed on acid-free, 250-year-life paper.
Manufactured in the United States of America.

Contents:

About the Author

Dan MacLeod has served as a consultant to employers, trade associations and unions for 30 years, and has developed innovative corporate programs that have resulted in thousands of ergonomic improvements and savings of millions of dollars. His record of experience is significant:

- Pioneered workplace ergonomics programs in auto plants in the late 1970s. Wrote the first lay language booklet on ergonomics in 1982, which was aimed at the auto industry and general metal working operations.

- Retained by the American Meat Institute and major meatpacking companies in the 1980s to develop industry-wide programs that have successfully reduced CTDs in this high-risk industry. Represented the industry in working with OSHA to develop the *Ergonomics Guidelines for the Meatpacking Industry*.

- Comprehensive experience in multiple industries, ranging from the office environment and hospitals to underground mining and steel mills.

- Conducted evaluations of work areas in more than 1000 different workplaces, involving tens of thousands of separate tasks.

Dan is a Certified Professional Ergonomist (CPE) and holds master's degrees in both occupational health and industrial relations. For more information, see www.danmacleod.com

Other Books by the Same Author

The Ergonomics Kit for General Industry (New York: Lewis Publications, 1999.)

The Office Ergonomics Kit (New York: Lewis Publications, 1998)

Acknowledgements

Content — Everything in this book is based on what I have learned working on-site in plants and with client organizations. In the process, I owe much to many people throughout the years, including the following very bright and helpful people: Eric Kennedy, Wayne Adams, Elizabeth Damann, Rob Nerhood II, Joe Pallansch, Don Chaffin, Tom Armstrong, Rich Marklin, Don Bloswick, Stover Snook, Bob Bozzay, Devin Lackie, Rob Radwin, Bill Boyd, Don Wasserman, Lois Hensley, Don Hirsch, and Ken McCombs.

Eric Kennedy and Wayne Adams especially provided considerable support in the initial development of the section on quantitative methods.

Layout and Design — The author.

Illustrations — Mark Watson supplemented previous drawings by Mary Noyes and Tom Nynas.

A previous version of this book was developed by the author for the Ergonomics Group of Clayton Group Services.

To Bertrande

Part I
The Rules

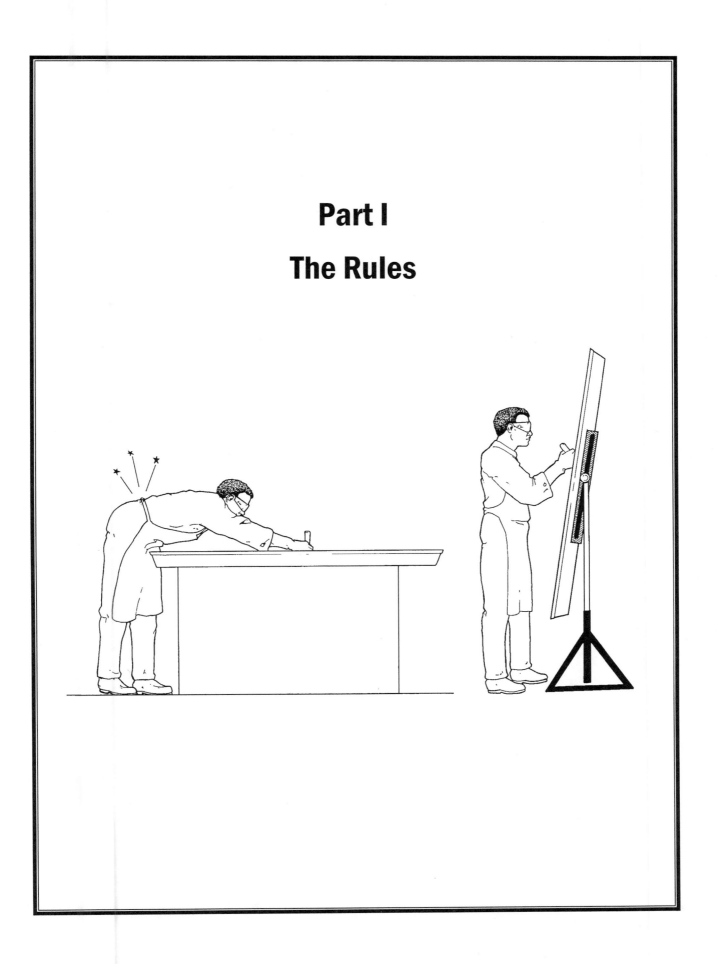

Introduction
Overview of the Field

Ergonomics is an interdisciplinary field of study that seeks to design tools, equipment, and tasks to optimize human capabilities. In this context, tools, equipment, and tasks are broadly defined. A *tool* might range from a simple hand tool, to a written set of directions, to an entire organizational system. *Equipment* includes factory production lines, household appliances, and sports paraphernalia. A *task* could be either a physical or mental activity; and it could be done as a job, a household chore, or a leisure-time pursuit.

The field seeks to optimize the interrelationship between the human and a system, whether that system involves a simple hand tool or an entire production line. Whenever one designs a more-effective interface between a human and a tool or task, that is ergonomics.

Targets for ergonomic improvement are divided into two broad categories: (1) physical issues, such as reaches and exertion, and (2) cognitive issues, such as mental overload and confusing displays. Synonyms for the field include man–machine systems, human–system interface, and human factors engineering.

Popular phrases that describe the field include:

Fit the task to the person — Probably the most useful phrase that describes the field is: "fit the task to the person, not the person to the task." At a basic level, this phrase asks whenever designing a tool or planning a task, "how does the human fit in?" At a more advanced level, the phrase means studying human beings and human behavior — anatomy, physiology, and psychology — and then designing tasks to fit these human requirements. The goal is to take conscious advantage of unique human capabilities when designing tools and equipment, and to counteract human weaknesses and frailties.

Work smarter, not harder — A time-worn phrase that many people aspire to is "work smarter, not harder." Normally, how one actually goes about doing so is left unstated. Significantly, ergonomics provides a *method* for finding smarter ways of working. It prescribes the principles and techniques by which people can improve ways to work.

User-friendliness. The term *user-friendly* is synonymous with ergonomics. Anything that can be described as user-friendly can also be said to be ergonomic; unfriendly items are not ergonomic.

User-friendly means that things are easy to understand and apply, that mistakes are reduced, and that the human is treated well in the process. This term first became popular in conjunction with computer software, but the concept can be expanded into every aspect of life, whether at home or on the job. One can refer to user-friendly tools and equipment, user-friendly offices and production facilities, or user-friendly highway systems and shopping centers (or "ergonomic" highway systems and shopping centers).

The rules of work — Finally, it is instructive to know that the term *ergonomics* was coined from the Greek words *ergon* (meaning "work") and *nomos* (meaning "rules"). Therefore, the literal meaning is "the rules of work," which provides good appeal, since everyone can benefit from knowing these rules.

Ergonomics provides a set of conceptual guideposts for adapting workplaces, products, and services to fit human needs. It offers a strategy for engineering design and a philosophy for good management, all with the underlying goal of improving the fit between humans and workplace activities.

- - -

All of this is ergonomics. It is a comprehensive concept that addresses the very core of work, whether for productive labor, for household chores, or even for leisure activities.

The focus of this book concerns the physical issues of the workplace in general industry and, as such, highlights the design of tools and workstations. However, it is crucial to keep in mind that the same concepts can be applied to consumer product design, and that ergonomics also encompasses cognitive factors.

The Human Factor

There are three basic human factors that must be accounted for in the design of the workplace: (1) people are different, (2) people have limitations, and (3) people have certain expectations and predictable responses to given situations. If these factors are ignored, design consequences can be costly, both financially and in terms of human discomfort and performance.

People are different — People come in different shapes and sizes. Some are tall, others short; some are young, some old. Yet, the workplace is often designed based on the premise that "one size fits all." The result is often that one size fits only a small portion of the population. Large segments of the workforce may work in sub-optimal conditions.

Older workers may not be as robust, agile, or keen sighted as younger workers. Older people provide great value because of their experience and capabilities, but we seldom take their special needs into account when thinking about workplace design.

Few people are completely able-bodied in every aspect of their lives. Many people have certain disabilities. Neglecting these differences in people means that otherwise capable people might not be allowed to work at all.

People have limitations — The physical and mental limits that confront humans are numerous. On the most basic level, there are overall limitations based on stature. For example, despite individual differences, there are limits to how far even the tallest person can easily reach. Nonetheless, equipment and products are often designed as though everyone had infinitely long arms, with switches and controls located where no one can reach.

The human body does not tolerate staying in awkward postures for a long time, such as with the arms outstretched or crouched in a stooped position. Further, the upper limbs cannot tolerate excessive motions without injury, nor the lower back excessive lifting, bending, or twisting. Pain results and work performance drops.

Mental capabilities — reacting to and processing information — can also become overloaded. There is a limit to how much mental activity people can perform simultaneously or for an extended period of time. Mistakes, errors, and incorrect decisions result when those limits are exceeded.

People have predictable reactions — People learn during the course of their lives to associate certain actions with certain signals, such as stopping at red lights or flipping switches up to turn on lights. By taking these predictable human reactions into account, the design of the workplace can be improved and the operation of machines and equipment can be made more user-friendly. When these expectations are violated — for example, by orienting light switches sideways — mistakes can easily result.

In addition to expectations concerning switches and control buttons, people respond in other predictable ways. The cognitive processes — how humans think, make decisions, and react — can also be predicted. As a final example, depending upon the task being performed, people can be predictably stressed, bored, or stimulated. The lessons learned by studying these predictable responses can be integrated into good design.

History

In a certain way, humans have been doing ergonomics for thousands of years. Indeed, modifying surroundings to fit capabilities and limitations is a defining trait of the species.

It is often helpful to view ergonomics from this perspective. It helps make the field more accessible and it lends itself to identifying inexpensive, low-tech improvements.

There are clear distinctions, however, between "old-fashioned" ergonomics and the practices of today. In the past, problem solving was haphazard and impulsive. Today, it is possible to be systematic, using a broad database of knowledge and applying analytic techniques, measurements, and the scientific method.

The birth of ergonomics as we know it today occurred in the World War II period, when scientists and engineers began to consciously study human capabilities and limitations with the goal of improving the design of military equipment, in particular aircraft.

In Great Britain, these classically trained scientists coined the term *ergonomics* for this application. In the United States, the term *human factors* was used, and in Germany the term was *arbeits-physiologie (work physiology)*. In subsequent years, the term *ergonomics* was generally adopted in Europe and recently has become much more widely used in the United States as an alternative to *human factors*.

To this day, more ergonomics studies have been performed on the aircraft cockpit than on any other single workstation. Indeed, the aerospace industry is the source of much of the data and concepts that are now being applied in industry. The role of aircraft design in the birth of the modern field of study accounts for the dual aspect of ergonomics in addressing both physical and cognitive issues. The classic questions posed about the cockpit were:

- <u>Cognitive</u> — Are the dials and controls understandable to a pilot? Do they perform according to standard expectations?
- <u>Physical</u> — Are all the controls within reach of every pilot? Is the operation of all manual cranks (such as for landing gear) within the strength capabilities of all pilots?

Ergonomics "versus" Human Factors

There has been a tendency in the United States to refer to cognitive design issues as "human factors" and physical design issues as "ergonomics." These two terms, however, are in fact synonymous. The more appropriate terminology is "cognitive ergonomics" and "physical ergonomics" (or "cognitive human factors" and "physical human factors").

In the United States considerable debate has taken place within the profession on this terminology. To be sure, making a firm distinction between the two branches is sometimes useful and professional qualifications tend to split along these lines. But in the main, the two terms should be synonymous. To think otherwise is to split the human being into a traditional "body" versus "mind" dichotomy, which is, in effect, not ergonomic. From a design standpoint physical and cognitive issues both must be taken into account.

The Scope of Ergonomics

In the same manner that a chemist can view the entire world in terms of chemistry, the ergonomist can view virtually every human interaction as ergonomics. Anything that helps humans expand on capabilities or overcome limitations can be viewed as components of this field. This schematic summarizes the scope of ergonomics:

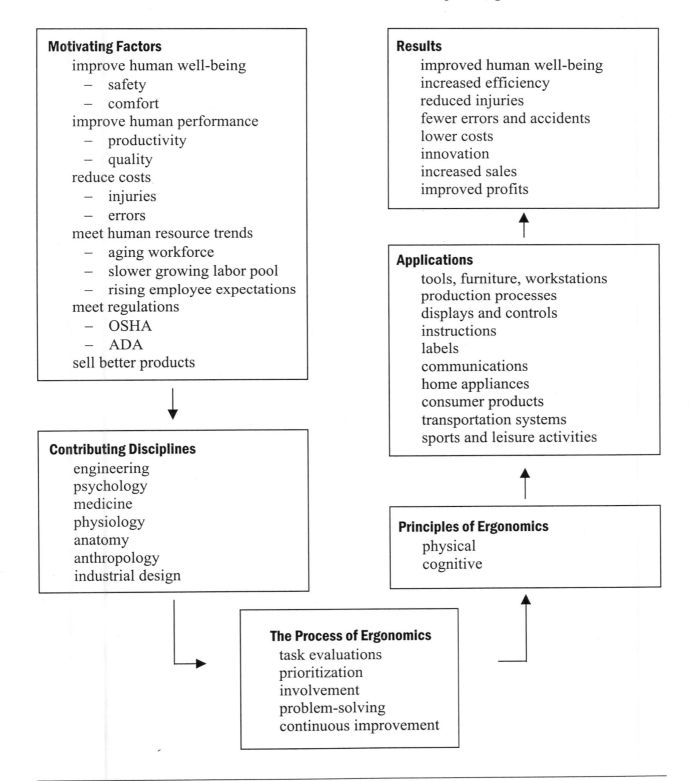

Motivating Factors
improve human well-being
- safety
- comfort

improve human performance
- productivity
- quality

reduce costs
- injuries
- errors

meet human resource trends
- aging workforce
- slower growing labor pool
- rising employee expectations

meet regulations
- OSHA
- ADA

sell better products

Contributing Disciplines
engineering
psychology
medicine
physiology
anatomy
anthropology
industrial design

The Process of Ergonomics
task evaluations
prioritization
involvement
problem-solving
continuous improvement

Principles of Ergonomics
physical
cognitive

Applications
tools, furniture, workstations
production processes
displays and controls
instructions
labels
communications
home appliances
consumer products
transportation systems
sports and leisure activities

Results
improved human well-being
increased efficiency
reduced injuries
fewer errors and accidents
lower costs
innovation
increased sales
improved profits

What Isn't Ergonomics

It sometimes provides insight to distinguish what ergonomics is not in order to recognize the scope of the field. The following statements offer some boundaries of what ergonomics is not:

- If a human is not part of the system or interaction, it is not ergonomics.
- If there is no underlying goal or impact in making improvements or affecting design, it is not ergonomics.
- If the concern is for the human alone, without regard to tools and systems, it is not ergonomics.

Application	What is ergonomics	What isn't ergonomics
Machines	Using a machine to reduce human labor	Improving purely internal mechanics of the machine
Automobiles	Easy access in and out of car door; comfortable seating; understandable dashboard; standardized and stereotyped controls; easy access to engine for repairs; air conditioning	Color and evenness of paint finish; fuel efficiency; engine type and horsepower
Air traffic control	Communications between pilots and air traffic controllers during takeoff and landing	Social relationships between pilots and air traffic controllers after work
Fishing	Comfortable seating (bass boat seat or soft soil at fishing hole); well-organized tackle box; easy access to landing net; sharp fillet knife	Skill in catching fish; enjoyment with fishing companions; pleasure from eating fresh fish
Hotel	Understandable directions to rooms; efficient check-in and check-out systems; standardized shower controls	Attitude of service staff; cable TV; wall decorations
Playing guitar	Wrist posture while compressing strings; noise exposure level	Happiness and self-fulfillment from playing music
Highway intersections	Clear signs and direction; standardized and obvious traffic patterns	Height of over passes; durability of road surface; land use issues
Personal self improvement	Training in work methods to perform tasks more efficiently	Improving self-esteem

People versus Machines

Many ergonomists have found it useful to contrast the strengths and weaknesses of both people and machines. Altho these statements sometimes seem self-evident, at other times they help put issues in perspective and give guidance to strategies and decision-making.

People...

Are at their best when they do a wide range of variable motions and tasks.

Are creative, can plan and invent.

Can react to unpredictable events, identify and choose options.

Are better at interpreting data.

But Machines...

Have difficulty doing complex motions and varied tasks.

Need programmed, pre-set responses.

Do specific functions within narrowly defined limits.

Are better at processing high volumes of data.

Machines Can Easily...

Do highly repetitive motions at high force and speed.

Detect small variations in routine tasks.

Perform accurately in endless repeated tasks.

Work in hostile environments (toxic, radioactive, extreme temperatures, etc.).

Have limitless design.

But People...

Are limited, wear out, and lose accuracy quickly.

Are easily distracted and make errors.

Are inconsistent, and need change in stimuli.

Are highly affected by the environment.

Have a fixed, limited design.

Cumulative Trauma Disorders (CTDs)

Much of the current interest and application for practical workplace ergonomics relate to the prevention of CTDs, a broad class of disorders that can approximately be defined as wear and tear from everyday tasks, whether at work, at home, or during leisure time activities. Synonyms include musculoskeletal disorders (MSDs) or repetitive strain injuries (RSIs). CTDs are referred to occasionally in the following materials, but are not described in depth because that information is readily available elsewhere. Also, ergonomics provides value far beyond the prevention of CTDs and these rules of work stand by themselves.

Principles of Ergonomics

Physical ergonomics can be itemized as a set of ten interrelated principles:

1. Work in neutral postures
2. Reduce excessive force
3. Keep everything in easy reach
4. Work at proper heights
5. Reduce excessive motions
6. Minimize fatigue and static load
7. Minimize pressure points
8. Provide clearance
9. Move, exercise, and stretch
10. Maintain a comfortable environment

Even tho many of these principles may appear simple and self-evident, they are routinely violated in the workplace. Furthermore, one should never underestimate the power of a few fundamental ideas applied systematically.

To be sure, there are other useful frameworks for categorizing the basic tenets of physical ergonomics, and clearly other valid terms and phrases. The intent here is to summarize the main areas of the field in a prescriptive manner that gives practical guidance in the workplace.

Furthermore, there is considerable overlap between these principles, depending on the task and the issues at hand, yet, each principle stands alone and serves to encapsulate a body of knowledge of the field of ergonomics. Taken as a whole, they represent a method for knowing what to look for, and finding smarter, friendlier ways to work.

It is crucial to understand the difference between a principle, which is universally true, and a rule of thumb, which can change depending upon the task and situation. It is more important to learn and use the underlying principles than it is to concentrate on the details of current prescriptions for specific problems, for several reasons:

- Specific rules can change and may be erroneous if applied in the wrong situation. Principles remain constant.
- The principles provide help for evaluating any task, whether at home, in the office environment, or in general industry.
- Technology, products available on the market, and best-known solutions all change with time. Understanding the fundamental principles helps in assessing these changes and keeping a practical perspective.
- The principles will remain the same, even when advancements in knowledge are made in the field of ergonomics.

The following pages contain descriptions of each of these principles, plus many common rules along with some examples of applications. Note the following:

- The principles refer to *sustained* work for the most part. Temporary departure may not necessarily involve any problem whatsoever.

- No numbers are used in describing these principles because none are needed to apply them successfully and to generate improvements. Likewise, there are no references to studies or data that provide the basis for the principles. However, subsequent chapters of this book review measurement techniques and key studies for the issues involved.

- In practical application in the workplace, it is helpful to categorize activities in two levels: (1) avoid extreme breaches of the principles, and (2) optimize the fit of the task to the individual.

- Each principle contains rules of thumb, examples, and tips; however, the point is to gain a clear understanding of the principles, not to memorize all these particulars, so that no matter what type of task is being reviewed, it is possible to recognize issues. Subsequently, if there is a specific task in mind, the particulars can be reviewed for ideas and guidance.

- As with any system, changing one part of a task may affect the rest. It is important to take a holistic view and attempt to anticipate the effects of any single change.

- Many of the examples in the following materials deliberately promote a low-cost, low-tech approach to problem solving.

- The illustrations show examples, but the concepts obviously go far beyond those shown. Part of the challenge — and satisfaction — of modern workplace ergonomics is thinking of creative ways to apply a solution from one application to another, even in a markedly different context.

Principle 1

Work in Neutral Postures

Neutral posture is the optimal position of each joint that provides the most strength, the most control over movements, and the least physical stress on the joint and surrounding tissue. In general, this position is near the midpoint of the full range of motions; that is, the position where the muscles surrounding a joint are equally balanced and relaxed.

There are important exceptions to this rule of midpoints. Examples are the posture of the arm, which is affected by gravity, and the knee, which functions well near its extreme extended position.

Key postures of practical relevance in the workplace are:

1. The back with its natural "S-curve" intact
2. The neck in its proper alignment
3. The elbows held naturally at the sides of the body and the shoulders relaxed
4. The wrists in line with the forearm

In understanding these postures, and especially in interpreting the following illustrations, it must be highlighted that the neutral posture for each joint is not necessarily an extremely precise position; rather, it involves a small range of positions. Furthermore, one should not conclude that there is one single best posture that should be maintained for the entire day: The body needs to change and shift occasionally (see the subsequent principles regarding static postures and movement).

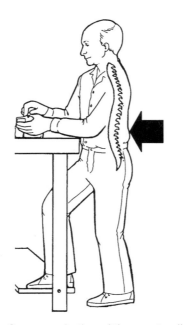

Keep the S-curve, whether sitting or standing

Maintain the "S-curve" of the spine — The spinal column is curved roughly in the shape of an "S." It is important to maintain this natural S-curve to prevent chronic back injuries and to optimize the working posture. For the lower back, this involves maintaining some degree of *lordosis*, that is, a slight "sway back," whether sitting or standing. Bending forward or otherwise flattening the slight sway back (*kyphosis*) puts pressure on the sensitive discs of the lower back, which can ultimately lead to severe back injury. Twisting of the back is similarly a key issue.

Correct alignment of the spine is facilitated by maintaining a semi-crouch posture, keeping the knees slightly bent. Positions that promote working in this posture include:

- Using a foot rest when standing
- Leaning back somewhat when sitting
- Tilting the chair seat pan down slightly
- Having good lumbar support

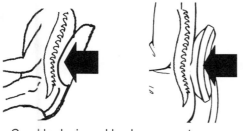

Good lordosis and lumbar support.

"C-curve" with inadequate lumbar support.

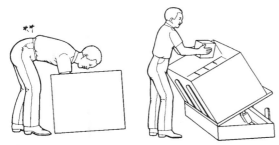

"Inverted V-curve" improved with use of a tilter.

Bent neck (and wrists) improved by tilting the machine.

Note in particular that the neutral posture of the back is not when standing straight upright like a soldier at attention. Likewise, it is not in neutral when sitting up straight with a 90° angle between the back and the thighs. Standing straight upright tends to exaggerate the curve a bit. Sitting tends to flatten the curve, especially if there is inadequate lumbar support.

Working in this semi-crouched posture with the correct amount of lordosis is optimal. However, in practice, priority should normally be given to extreme abuses, such as working hunched over (the "C-curve") or even worse, bending at the waist to reach the floor (the "inverted V-curve"). It should also be emphasized that often the incorrect posture is not a merely a bad habit of the person; rather. it is required by the poor design of the equipment and furniture. In particular, the old adage of "lift with your legs, and not with your back" cannot be adhered to in some circumstances, such as lifting out of a large box or tub.

Keep the neck in its proper alignment — The neck is obviously part of the spinal column and thus subject to the same requirements as the lower back. Chronic damage to the discs in the neck can lead to injuries as severe as those affecting the lower back.

The neutral posture of the neck is fairly self-evident; namely, it should not be bent or twisted. The technical terms again are lordosis for the proper concave curve (*excessive* lordosis, such as looking upward for an extended period, is a problem) and kyphosis for the stressful convex bend forward.

Keep elbows in and shoulders relaxed — Understanding the neutral posture of the arms is also quite self-evident. The elbows should be held comfortably at the side of the body; the shoulders should be relaxed and not hunched. Working with the elbows winged out can add strain to the shoulders and cause fatigue and discomfort, interfering with people's ability to do their jobs well and contributing to long-term injury of the shoulder.

If it were not for the effects of gravity, the neutral posture of the arms *probably would* be winged out — at least to some degree — because that is the midpoint of the range of motions. Consequently, if proper armrests are provided, there *may* be no

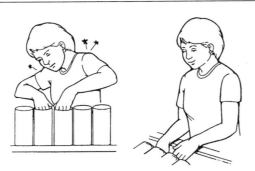

Changing the orientation of the product enables the elbows to be dropped to the sides and shoulders relaxed.

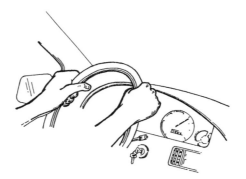

Keep the hand in the same plane as the forearm.

Ideally, the hands should be slanted slightly in and forward as well.

The contoured keyboard has been designed to keep the wrists in neutral while typing.

problem with having the arms winged out somewhat.

Keep wrists in neutral — The neutral wrist posture is much less intuitive to understand. The hand should be (a) in the same plane as the forearm and (b) angled somewhat in, more or less as if holding the steering wheel of a car at the 10 and 2 o'clock position. One can also directly check the neutral posture by dangling the arms at one's sides and observing the position of the wrists.

Note in particular that the neutral posture of the wrist is *not* at right angles like it is when holding a bouquet of flowers or playing the piano. It is halfway between those positions, tilted slightly forward.

Again, it should be emphasized that the priority is to insure that no one is working with a severely bent wrist. Optimizing the wrist posture is a secondary step.

General Design Implications

There has been a tendency in the past to think about good human posture in terms of right angles. For example, there's an old (and inaccurate) rule of thumb called "90° – 90° – 90°" that recommends sitting and working using right angles for the body. Furthermore, equipment is typically constructed with right angles (desks, workbenches, doors, windows, cabinets, etc.) both because it is conventional and because it is generally less expensive to build that way. These ever-present right angles tend to reinforce the notion that posture should be in 90° angles.

The human body, however, is not rectangular and does *not* necessarily work best at 90° angles. The neutral postures of the various joints of the body are much more irregular than that. Consequently, there is a need to learn this new set of rules about working postures.

Workplace Surveys

Based merely on this one principle regarding neutral postures, it is possible to do a credible workplace survey and make meaningful improvements. The steps are to review each job for awkward and contorted postures — bent or twisted back, bent neck, elbows away from the body, and bent wrists — then follow the implications thru to their logical conclusions about how the tools, equipment, or tasks could be changed.

Postures can be improved by various strategies:

- changing heights and reaches
- tilting equipment
- modifying equipment layouts
- designing pistol grip or modular grip tools

Examples

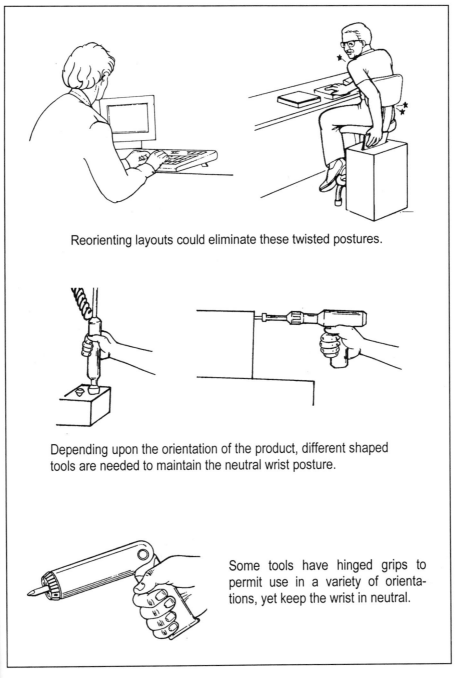

Reorienting layouts could eliminate these twisted postures.

Depending upon the orientation of the product, different shaped tools are needed to maintain the neutral wrist posture.

Some tools have hinged grips to permit use in a variety of orientations, yet keep the wrist in neutral.

Both this unconventional knife and chair were designed based on a sound understanding of neutral postures. Although the applications for each is limited, the concepts provide memorable, graphic examples that help make the educational point.

Often, many adaptations in a piece of equipment, or alternative technologies, can help eliminate contorted postures. In the case of a telephone, options for improvement include headsets, speaker phones, phone cradles, and phone cushions.

There is no reason not to lean back in a chair, as long as there is good lumber support to help maintain the S-curve. In fact, leaning back like this is better than sitting up at right angles.

Principle 2

Reduce Excessive Force

Countless tools and machines thru the ages have served to reduce exertion. Indeed, leveraging human capabilities to increase force is one of the hallmarks of human progress. The search for yet further ways is one of the defining areas of contemporary workplace ergonomics.

Although the terms force and exertion are used interchangeably in everyday language, technically the two are different. In the context of workplace ergonomics, *exertion* is the tension produced by muscles and transmitted through tendons to produce force. *Force* is the externally observable result of a specific movement or exertion.

This distinction is important for two reasons. First, people of different physical sizes may produce the same *external* force to accomplish a task, but may experience different *internal* muscle tension. It is easier for a fit 200-lb. person to push a 50-lb. load than it is for a 98-lb. weakling. Second, there is the issue of mechanical advantage, or perhaps more appropriately, mechanical *dis*advantage. Due to the inherent structure of the musculoskeletal system, exertion levels are often many times larger than the resulting external forces. Pushing a 50-lb. load can result in loads several times that amount on the elbows and shoulders.

Excessive force can overload the muscles, creating fatigue and potential for injury. Furthermore, applying excessive force to perform a task can slow down the effort and interfere with the ability to perform the task well. Consequently, almost anything that can minimize the exertion required for the task can make it easier and typically faster to perform.

Force is a term that is relative to the affected muscles. For example, the amount of force needed to depress an individual key on a keyboard is a crucial issue for the hands, but would obviously be trivial for the whole torso or legs. Thus, in workplace ergonomics "force" is often used in instances where it would not normally be used in daily life, such as referring to pinch forces used to assemble minuscule products like heart pacemakers or other medical products.

Duration is also an important consideration in dealing with force. A force that needs to be sustained for minutes has quite a different implication from that needed to be applied for only a second. The effect of duration is important, but one that is more appropriately addressed in the context of static load, which will be discussed later as a separate principle.

With these background comments in mind, the following are a series of rules for the design of work. To reemphasize, these rules are *highly task dependent,* and a rule that can be helpful in one situation may not apply in another. Taken as a whole, these rules do provide excellent guidance in the process of innovation for problem solving.

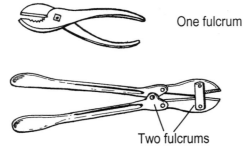

Ball-handled (as well as pistol grip) screwdrivers increase the mechanical leverage of the hand.

One fulcrum

Two fulcrums

Compound levers such as in this bolt cutter provide tremendous mechanical advantage, even when incorporated into small tools used to squeeze tiny parts.

Levers

Increase leverage — Levers are as old as the earliest humans. Levers are usually thought of in terms of lifting large, heavy objects, but can be applied in virtually any circumstance, such as increasing the diameter of a screwdriver handle.

There are many opportunities in industry for providing levers for the hands and fingers as well as heavy loads on the back. One sees, for example, even in high-tech facilities, assemblers pressing together parts with the force of their hands, instances where simple presses and other types of levers could be employed to great effect and at low cost.

A particularly underutilized type of lever is the compound lever, an age-old technique that multiplies the effect of levers. A compound lever can produce the same force as a simple lever but using only one-fourth the length and with a corresponding decrease in distance that the lever arm needs to be moved.

Compound levers have traditionally been used where extremely high levels of force must be generated, such as in bolt cutters. However, the same concept can be applied for other tasks, even tweezers-size compound levers to reduce some pinch forces.

Reduce tool length — If force is applied to a tool such as a paint scraper, then the shorter the length of the tool, the less effort is required to produce the needed force on the working end. Some tools, like scrapers, can easily be cut down. Other stubby tools are available on the market for this reason.

Bear in mind it can sometimes be helpful for hand tools to be long enough to enable use of both hands for extra force. The main point, however, is that the longer the distance between the hands and the working tip of the tool where the force needs to be applied, the greater the exertion required because of the mechanical *dis*advantage. Consequently, cutting the length down (or choking up on a tool) increases the mechanical advantage.

The traditional scoop design (above) is off balance; that is, the hand is a fulcrum and the load is far to one side. In the unconventional design (below) the center of gravity is below the hand and the force on the wrist is reduced.

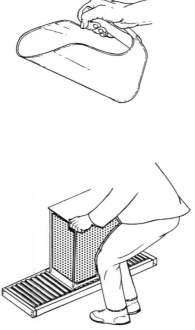

Even short lengths of conveyors can help.

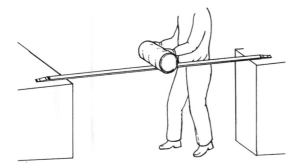

Skid bars are an inexpensive way to reduce the amount of lifting and carrying.

Improve tool balance — A tool that is said to have good balance means that the center of gravity of the tool is close to the wrist. It is helpful to think of the tool as a lever, with the hand as a fulcrum.

The farther the weight is from the center, the greater the load on the wrist. In biomechanical terms, the *moment* on the wrist increases.

Slides and Rollers

Use conveyors — Traditionally, waist-high conveyors have served a useful role in eliminating the need to carry heavy loads. Useful expansions of the conveyor technique include:

- Floor mounted conveyors
- Adjustable height conveyors (mounted for example on scissors lifts)
- Flex or snake conveyors
- Short lengths of roller conveyors (for example, six inches to permit a heavy part to be slid into a machine tool rather than lifted)
- Ball or omni-directional conveyors

Note that there are several ways to build gates into the conveyors to permit egress, including drawbridges and slide gates.

Use skids — Conventional techniques for reducing lifting have centered on hoists of various types. However, skids, slides, and rollers are often cheaper to install, faster to use, and more likely for people to use. (Humans will often instinctively drag loads when possible rather than lift. In contrast, people will commonly lift a load manually rather than go out of their way to fetch a hoist.) Even a skid as short as a few inches can be helpful.

Use funnels — A variation on this theme is to use funnels, chutes, or guides to help locate a heavy object into its position. The guide can support the weight as well as minimize the amount of fumbling to place the object in its proper orientation.

Improve layouts — The need to lift or manhandle equipment can be reduced or even eliminated by optimizing the arrangements of work areas. Examples include equalizing work surface heights, eliminating lips and barriers, and simply reducing

Poor layout (especially uneven heights) often causes forceful, wasted motions.

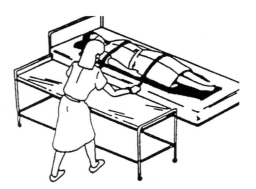

Air-powered patient transfer. The patient lies on a type of air mattress that has holes on the bottom. Air is pumped into the mattress, which lifts it and the patient slightly off the bed, reducing the friction almost entirely.

To reduce force, increase wheel size, add handles, and improve flooring.

reaches and distances that items need to be moved. Moreover, if work surfaces are equalized, then it becomes possible to slide products and materials.

Make surfaces slippery — If loads need to be slid along a surface, there are several options for reducing the surface friction:

- Construct the surface from dimpled steel or low-friction plastic.

- Use air bearings or air tables, which use pressurized air forced out thru small holes in the surface to create a near frictionless film of air. Air tables are used routinely in the paper and wood products industry, but unfortunately almost nowhere else. New applications include pallet jacks, lifters for heavy machinery, and hospital patient transfers.

- Imbed ball rollers in the surface.

- Add lubricant or even sprinkle corn meal on surfaces.

Carts

Pulling on a cart or other similar load is prevalent in many types of operations. Obviously, using carts is generally better than carrying and manhandling boxes and supplies. However, heavily loaded carts can create high force.

Generally, it is better to push loads than to pull them. However, pulling is often better when going over a curb, a gap (such as with an elevator), or any other type of obstruction. Pulling helps lift the cart upward and over the barrier, whereas pushing tends to force the cartwheels down into the barrier.

The goal, however, is to reduce the force required altogether:

- Use larger wheels with good-quality bearings.
- Keep the floor and wheels in good repair.
- Provide good handles.
- Use a power tugger if necessary.
- Use rails recessed into floors for routinely used, extra-heavy carts.
- Overhead monorails have like advantages.

Secure the Equipment

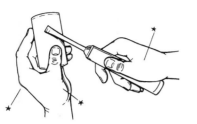

Using the hand as a fixture increases exertion and wastes effort.

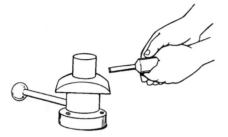

Fixtures reduce exertion

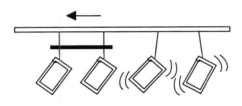

Backstop (or guide rail) to secure parts hung on overhead conveyors.

Use fixtures — The negative rule is clearer: *Don't use the hand as a fixture.* It is common to see people holding onto an object with one hand and working on it with the other, thus exerting against themselves. If one simply fixtures the object, it can become much easier to work on.

Furthermore, the fixture frees up both hands to do the actual work, rather than simply holding onto the object. Put another way, using the hand as a fixture is non-value-added activity.

Using a vise mounted to a workbench is one of the most common examples of a fixture. Placing a board in a vise makes it much easier to saw on rather than just holding it on the bench (and especially easier than trying to saw a board while holding it in the air).

Use backstops — A backstop is a very simple fixture, such as a heavy weight or a lip secured to a workbench used to prevent a product from being pushed around while it is being worked on.

A similar application addresses the type of overhead conveyors from which products are suspended and which typically swing, and which consequently often require using one hand to stabilize the product while the other hand performs the work. Attaching a bar of some type alongside the overhead conveyor can often help steady the product and reduce manual exertion.

Reduce Resistance

Heat malleable products — If plastic or a similar product needs to be manipulated or fitted, it is often helpful to heat the material to make it more malleable. Heat lamps and hand-held heat guns (like a hair dryer) have been successfully used in the workplace for this purpose.

Reduce resistance of activators — It is common for the tension of hand-activated devices such as triggers, control buttons, or levers to be set unnecessarily high. Force can often be reduced by changing springs, the mechanics of the controls, or by maintenance. In some cases, power can be added to reduce manual exertion, such as with power steering and power brakes on a car.

Proximity switches and photoelectric triggers eliminate force.

Hoses and cords — A sizable percentage of tools are connected with rather heavy cords and hoses. Air-powered tools in particular are connected with hoses that are often stiff and heavy, especially when weighted down with large metal fittings. Shop jargon refers to the wasted effort involved in manipulating the tool as "fighting a hose." Improvements include suspending the hose from a counterbalance, plus using swivel couplings.

Improve Grips

Grip design is a key issue and the subject of many of the rules of work. *Grips* refer to the handle on a tool as well as other applications such as on a pushcart or a toolbox.

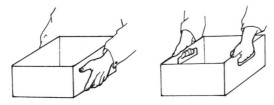

Handholds enable use of a power grip and thus permit more work for less effort.

Use power grips — One can exert more force with a full hand grip (a *power grip*) than with the fingers alone (a *pinch grip*). So, if the task requires strength, provide tools and equipment that make use of a power grip.

A good example is carrying a box or tote — boxes with handholds take less exertion to carry. Consequently, with a good grip, one can accomplish the same task with less effort.

Use appropriate grip size — The optimal size for a power grip is roughly that which permits the thumb and the forefinger to overlap slightly. If the grip is significantly larger or smaller than that, more exertion is needed to accomplish the same work. Creatively designed hand tools like some pliers and gripping tools, play to this advantage.

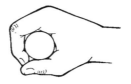

To generate the most grasping force, design objects to a size that permits the thumb and forefinger to overlap slightly.

The size of small grips can easily be increased with wraps and covers of various sorts. Indeed, many people routinely use tape to increase the size of small grips (in contrast, it is difficult to reduce the size of excessively large grips).

The implication is that, ideally, different-sized grips should be made available for persons with different-sized hands. In some cases, it may also be appropriate to use adjustable or custom-made grips. This concept is not as far-fetched as it may seem at first because traditionally — that is, before mass production — tools were routinely custom made to fit the size of individuals' hands.

It takes more effort to generate a given force if a tool grip is excessively small (or excessively large).

As a reminder, this grip size is generally valid only for tasks that require power. If precision is required, or other aspects of the task come into play, this rule of thumb may not be appropriate.

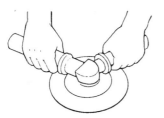

Two hands can be better than one.

Friction surfaces on grips reduce force.

Let tools do the heavy work.

Use two-handed tools — Adding a second handle to a tool can halve the exertion required when one hand alone holds a tool. Having two handles also permits greater control and more accuracy of the tool. "Tommy gun" grips are a version of this concept used for pressure hoses. Note that a prerequisite for a two-handled tool is that the product being worked on must be secured.

Increase contact friction — If a grip is slippery, the user must exert more to accomplish the task. Thus, a common approach to reducing exertion is to cover the grip with a material that provides higher friction. Pencil grips are an example, as are the thin rubber wraps on dental tools. Employees often wrap tools with tape or similar material in an effort to increase the friction (as well as to increase size).

Apply custom-molded grips — Plastic wraps that can be custom-molded to the individual user's hand are available on the market. The process is simple and involves wrapping the grip in the plastic, heating it, and then gripping the tool. While the material cools, it forms to fit the exact shape of the hand. This approach is suitable if the grip is only held in one position; if the grip is rotated or flipped, the custom groves can obviously create problems.

Use a collar — In cases where the force applied is coaxial to the grip, providing a collar or stop on one or both ends of a grip can reduce grasping force. Without the collar, the hand must squeeze harder to accomplish the same task.

Mechanize

Use power tools, machines, and cylinders — Perhaps the most obvious way to reduce force is to completely mechanize the activity, which has been the core of industrial progress over the past two centuries: steam engines, electrical equipment, and so on. Power cylinders are particularly useful, whether powered by pressurized air, oil, or water (the last being an underused technology that has great advantages in the food processing industry or other areas where oil leaks would present a problem).

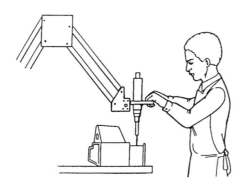

Mechanical assist (and articulated arm)

Vacuum hoists provide the advantage of instantaneous coupling.

Tool counterbalances reduce arm exertion.

Use mechanical assists — One way to reduce exertion involves mechanical assists, which are tools or pieces of equipment that provide force, leaving the human to guide the process. Assists of this type often involve use of an articulated arm; that is, an arm that has joints.

Mechanical assists can be powered, or simply be a piece of equipment rigid enough to absorb the force of the task. They can be elaborate devices or sometimes simple, inexpensive contraptions.

A host of new lifting equipment has become available in recent years that can be used in thousands of applications.

- Mechanical arms
- Adjustable-height carts
- Vacuum hoists
- Hydraulic lift and tilt tables
- Dumpers of various types
- Dollies of all kinds

Note that it is possible to increase the size of the loads once these kinds of lifting aids are used, which can increase efficiency.

Design for mechanical help — If equipment must be moved regularly, provide a design that facilitates use of mechanical devices. An example is equipment that is switched out for different production runs or removed for cleaning; this equipment can be fitted with brackets to permit use of a fork lift.

Use counterbalances — Many loads that are manipulated by the arm can be counterbalanced to make them virtually weightless. It is common for heavy tools to be suspended.

Techniques include:

- Suspension, usually with some type of spring, as is commonly used to support tools.

- A weight in combination with a pulley or some type of lever and fulcrum.

- A spring-loaded cylinder, such as those found in cars to help support the weight of trunk lids and engine hoods.

A curling bar places muscles and wrists in a better posture than a straight bar does and it enables lifting heavier weights.

The two-handed scythe is a noteworthy nineteenth century ergonomics device that reduced exertion by (a) enabling use of the larger muscle groups in the torso and (b) taking advantage of the mechanical structure of the upper body.

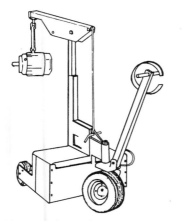

Mobile hoist for maintenance personnel

Body Position

Design for neutral posture — One advantage of designing for neutral postures is that, in general, the muscles are in their optimal position for strength when in their neutral position. This concept is quite intuitive: It is much easier to exert muscles in a natural posture than in an awkward one. Thus, by designing for neutral postures one can do more work for less effort.

There are some exceptions to this general rule, such as when a non-neutral posture permits use of additional muscle groups (for example, pulling a load with the arm fully extended and using the larger muscles in the torso and leg).

Enable use of larger muscle groups — Equipment and tasks can be set up in a variety of ways to permit use of additional, larger muscles.

- It is often helpful to provide a high angle of attack for pushing down on an object. For example, it may be possible to add a shelf below the surface of a workbench where products can be placed for downward pressure. This enables use of larger muscles in the torso, plus takes advantage of body weight.

- Generally, it is better to exert fore and aft, rather than sideways. It is common in industry to work on a bench where merely reorienting the product or process 90° can help reduce exertion.

- A related concept is using gravity to slide materials down, rather than lift them up. (Simplistic as this sounds, it is all too often overlooked as a solution.)

- It is sometimes possible to employ use of the large leg muscles. The classic example is the lifting technique of using the legs rather than the back.

Design for maintenance — Maintenance personnel often perform heavy exertion while in contorted postures because they frequently work in difficult-to-access locations. Methods include improving the access to needed locations, using various types of mobile hoists, and designing for mechanical help.

Arm brace on a "reacher."

Transfer the Load

Use arm braces — If a tool is used to hold a heavy load, then it may be possible to add a forearm brace. The classic examples are (a) the "reachers" used in old time grocery stores and (b) high-powered sling-shots.

Backpack belt — Another example of this general technique is the backpack belt, which shifts the load off of the spinal column onto the hips and legs. One wonders if this concept could be applied elsewhere, e.g. self-contained breathing tanks, holsters for heavy cords or tools, etc.

Training and Technique

Over reliance on the simplified instruction to "lift with the legs" has proven to be ineffective as a method for reducing back injuries. However, lifting technique can be important in instances where it is not feasible to provide ergonomic improvements.

- Keep the load as close to the body as possible.
- Keep the S-curve of the back when possible.
- Move the load smoothly; no jerking motions.
- Tilt the load up for a less extreme bend.

Workplace Surveys

With these basic rules and tips regarding force in mind, it is possible to survey a workplace and identify many issues and potential improvements. One merely assesses each task, looking for indications of force — pinching, grasping, pushing, pulling, lifting, and so on. Then, for each issue observed, one considers as many alternatives as possible.

In initial stages of an ergonomics assessment it is usually not important to ask "Is this a problem or not?" or "Is this an unsafe level of force or not?" If it is possible to make a simple, low-cost improvement to reduce some force, then the change can typically be made without further study. If the only possible improvement is expensive, or if there is some other extenuating circumstance, then it may be useful to determine if the loads are excessive. (See subsequent material on measurements and guidelines.)

It should also be mentioned that ergonomics cannot solve all problems for both technical and economic reasons. However, many problems *can* be solved, often in ways that are low cost, if these rules of work are applied with innovation and creativity.

Principle 3

Keep Everything in Easy Reach

An essential aspect of task design is keeping products, parts, and tools that are frequently needed within easy reach. Focusing on reaches provides value in several ways:

- Long reaches are often the underlying reason why people are found working in awkward, contorted postures.
- A long reach combined with lifting a load can increase exertion and multiply the load on the shoulders and lower back.
- Long reaches can be a source of wasted time; thus, they need to be addressed from an efficiency standpoint.
- Addressing reaches can lead to better use of valuable floor space and can reduce congestion.
- Finally, in some circumstances the ability to reach at all is the issue. For example, a pilot in a cockpit does not have the luxury of standing up, using a stepstool, or employing any of the other normal accommodations that people do when an item is beyond reach. The location of controls must simply be within everyone's grasp.

It should be noted that in many ways, this principle is redundant with that of posture. If the postures are in neutral, then the reaches will normally also be satisfactory. However, in the case of reaches, one evaluates the equipment and workstations themselves, whereas with posture, all the signs of problems come directly from observing individuals. Thus, evaluating both reaches and postures serves as a valuable way to double-check observations by viewing the task from different perspectives.

Moreover, the techniques for evaluating reaches permit evaluation of workstations where there are no employees at hand to observe directly. Thus, considering reaches can be helpful when planning a new workplace and design decisions are based on drawings.

General Rules

Maintain the reach envelope — The basic rule is to always keep in mind the *reach envelope*:

- Frequently used materials should be kept within the reach envelope of the full arm.
- Things that are almost constantly in use should be within the reach envelopes of the forearms.

Note that this envelope is a semi-circle, not the rectangle typically used for work surfaces. The envelope is sometimes also referred to as the *swing space* of the arm or forearm.

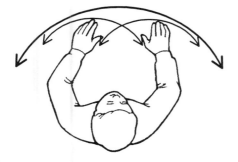

Reach envelope

Design for the short person — A useful rule of thumb is that reach should be established to accommodate smaller-statured people. The idea is: If shorter people can reach, so can everyone else (there is an opposite rule later regarding clearance).

Improve heights — There is an obvious interrelationship between heights and reaches. Thus, one way to reduce reaches is to optimize heights (see the following principle).

Rules for Worksurfaces

Rearrange — Keeping things in easy reach is not what anyone would call a hard concept to grasp. What *is* difficult is having the presence of mind to notice the reaching. Typically, long reaches are so habitual that individuals are unaware that they are doing so or that items could easily be moved closer. Even casual observers are commonly so used to seeing the task that they may not notice the reach as it occurs.

Once an unwarranted reach is identified, tho, making the needed change is often simply a matter of arranging the work area and moving things closer. At the computer workstation, for example, the idea of the reach envelope helps users identify instances where the keyboard, mouse, and phone are located too far away.

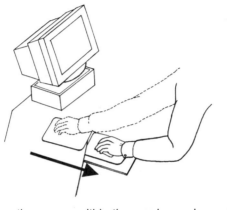

Keep the mouse within the reach envelope of the forearm.

Provide a place for everything — A predictable cause of long reaches is poor organization of space: tools placed randomly on a surface, spare materials and personal items in the way, and a general lack of order. By thinking thru layouts and providing a specific place for everything, more workspace can be created with fewer reaches.

The epitome of good design in this instance is the house trailer or recreation vehicle, where a surprising number of items can be stored in a small area by using some creativity. These are not difficult concepts, but they do not just happen by themselves.

Operational controls are often located beyond easy reach and can be fixed with low cost.

Relocate controls — It is common for machine controls to be located excessively far away. Relocating the controls is a relatively easy and low-cost improvement. With some styles of controls, such as with some two-hand safety buttons, it is occasionally necessary to keep the controls at a distance. But with other styles, the safety issues can be met, plus the long reaches reduced.

Reduce worksurface size — Tradition holds that the larger the worksurface the better, often implying higher social status, at least among executives. But large work surfaces invite and create long reaches. Thus, in some cases it is helpful to reduce the size of a given work surface. Smaller can actually be better.

For example, in light assembly operations, the tendency is to provide everyone with standard six-foot or eight-foot benches, but where the task actually requires a work area of only two-foot wide and one-foot deep. In such cases, reducing the size of workbenches provides an important side effect — picking up floor space and reducing congestion.

Furthermore, a small worksurface can be mounted on a pedestal. This further reduces congestion, plus makes it easier and cheaper to provide height adjustment.

Make cutouts — As another alternative, if there is a need for a large work surface area and there are problems with long reaches, it is possible to have a cutout made in the work surface. Cutouts provide a good way of reducing reaches while still allowing large work areas.

Use swing arms — Another way to bring items closer yet sparing work surface space is to use arms. Such arms can be adapted to hold tools, parts, instructions, fixtures, lights, telephones, and indeed almost any item normally found on a workbench. Lazy Susans have also been used successfully for this purpose.

Use reachers — A variety of hooks, rakes, poles, and grabbers can be used. If the load is heavy, the reachers can be modified as previously discussed regarding force.

Remove barriers — Many reaches are caused by barriers that can be eliminated or relocated.

Cutout

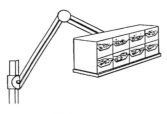

Swing arm

Remove barriers than create long reaches. Note also that long reaches like this create unnecessary exertion.

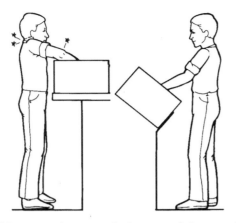

Tilting container stands is one of the most common, low-cost ways of reducing reaches.

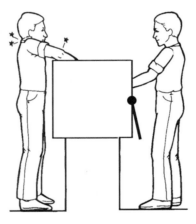

Container with drop down side

Reaching for Parts

A common problem in manufacturing is reaching for and replacing completed parts. Such long reaches can be improved in a number of ways.

Tilt — When working out of boxes it is possible to use tilt tables or stands, or even just prop up the box on one end. Tilted box stands can easily be fitted with hinges, cylinders, and even rollers to enable easy transfer to carts and conveyors.

Use drop down containers — Many styles of containers incorporate sides that can drop down or be removed. These designs permit ready access to materials while cutting down on the reach.

Use chutes — Parts can be fed directly to people by using chutes and hoppers. This approach is particularly helpful when there are multiple parts that cannot all be kept in boxes on the same work surface.

Use slider tracks — When parts are handed off from one person to the next, as is becoming more common, it is possible to use slider tracks to facilitate the transfer. This technique is useful when the workstations are further apart than an easy reach. Furthermore, a track mounted on a hinge can be used to slide parts across an aisle in a work cell (the hinge permits easy movement to allow passage).

Use smaller lot sizes — Large containers typically cause long reaches. Depending upon size of parts and volume used, it may be possible to use smaller lot sizes (as with Just-In-Time inventory control), which typically involves using smaller containers, thus shorter reaches.

Large Products or Equipment

Many large products and pieces of equipment require reaches that can be difficult to resolve. However, types of improvement are to:

Use shuttles — Shuttles are traditionally used to improve machine efficiency because the employee can load or remove parts from one part of the shuttle while the other part is in use. Fortuitously, the in-

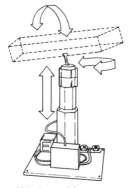

Work positioner

Trunnion

dexers also tend to reduce the reach. Shuttles and indexers and can be used for this purpose in a variety of applications.

Use work positioners — Products that are worked on for a period of time can be affixed to work positioners. These are fixtures that can be oriented to present the product to the employee in the optimal position. Work positioners can be powered and even programmed so that the product automatically shifts to the next position.

Use trunnions — Some types of large products can be attached to trunnions, which are big fixtures with pivot points that permit turning objects over into various positions.

Build subassemblies — When products are assembled on moving conveyor lines, many reaches are related to intractable items, such as the size of the conveyor, the rigidity of the position of the product and its pace, and the required space between products. However, assembling products off line yields many more possibilities for reducing reaches, such as by using trunnions and work positioners. An often-feasible approach is to build subassemblies off line, then fit the whole subassembly into the larger product (perhaps with the aid of a mechanical arm).

Training

Once improved equipment is put in place, proper training of operators must not be neglected. If adjustable devices are provided, then everyone must know how they work. Trainers should have operators try the devices and practice a while under supervision rather than simply providing oral — or sometimes even worse, written — instructions.

Training on how to change the tension in tool balancers can result in improved use.

Principle 4

Work at Proper Heights

A common workplace problem is a mismatch in heights between employees and the work that they are doing. This leads to poor postures and related fatigue, discomfort, and potential damage to soft tissue. Moreover, awkward heights quite often create unnecessarily harder work and decrease the ability to perform the task correctly.

Proper height depends on the nature of the task. Once again, this principle is often redundant with posture. If the postures are correct, then generally the heights are correct. However, exertion is also affected by height, and not always in correlation with neutral postures. Other task issues can also affect the best working height, so, in practice, identifying the best height for a workstation may not be clear-cut. Indeed, it may not be possible to create ideal heights and some judgments may need to be made to find the best compromise (this can also be true for the other principles and, in fact, for every type of engineering and design). Nonetheless, a number of rules provide guidance in optimizing heights of equipment.

General Rules

The first priority is to avoid extremes.

Generally, working at about elbow height is optimal.

Avoid extremes — Many times, when it is not practical to design every height to be optimal, it may be feasible at least to avoid the extremes; that is, avoid working below knee level or above the shoulders. For example, racks can often be modified by removing or blocking off the extremely high and low rungs. The same is true for carts, shelving, and other similar equipment. Fixed-height stands at about knee level can provide a low-cost way to avoid working at extremely low levels.

Design for elbow height — Generally, work is best done at about elbow height, whether sitting or standing. This is true for computer keyboards as well as other kinds of work in manufacturing and assembly.

Note that it is *the work itself* that should be at elbow height, *not necessarily the work surface*. For example, if unusually large products are being used, the heights of conveyors and other work surfaces should be adjusted accordingly. The issue is the *height of the task* being done, not the height of the work surface.

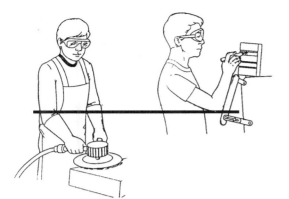

Heavier work may be best performed lower than elbow height; precision work higher.

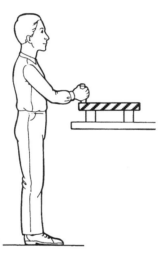

Platforms on the work surface can accommodate tall people; platforms on the floor, shorter people.

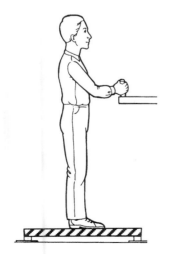

Consider the exceptions — The nature of the work also affects the proper height. Heavier work, requiring upper body strength, should be lower than elbow height. Lighter work, such as precision work or inspection tasks, should be higher.

Consequently, the nature of the task must be taken into account when designing proper heights. It is not always sufficient to look up a height dimension in a table of numbers or just to apply a common rule of thumb.

Adjustable heights

Because people vary in height, good design usually involves providing some sort of height adjustment. There are a variety of ways to meet this need.

Best: Change the work surface — When possible, the best approach is to adjust the height of the work surface itself. It is easiest when only one person uses a particular workstation. That workstation can then be adjusted once for that person, for example, by lengthening or shortening the legs of a workbench.

If several people use the same workstation, it becomes more difficult. Placing some sort of a simple riser or platform on the work surface can sometimes accommodate taller people.

A more elaborate, but increasingly common option is to use adjustable workbenches. These can either be adjusted manually by use of a crank or they can be powered and push-button controlled.

A huge variety of adjustable equipment has come onto the market in recent years. Each type has its advantages in specific situations, so with careful selection, good solutions can be found for many needs.

Second best: Stand on platforms — When working on machines, conveyor lines, and other large pieces of equipment, it usually is impossible to raise and lower the work. The alternative, then, is to raise and lower the operator. This concept has the disadvantage of creating congestion and even a potential tripping or falling hazard. Nonetheless, standing platforms may be the only option and they have worked exceedingly well in many facilities.

Platforms can be something brought over from a storage spot when needed, or even be a type that can be pulled out or flipped down from under the work-

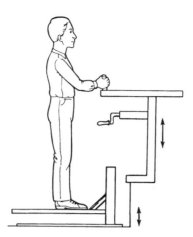

Powered adjustable worktables and platforms are becoming increasingly common.

Example of retrofitted tool extender.

Height differences in equipment can create unnecessary work.

station. Alternatively, the platform can be powered for ease of adjustment.

Set heights for tall people — If standing platforms are used, then the usual approach is to design for the tall person and provide platforms for shorter people. The rationale is that is it feasible to raise shorter people up, but difficult to lower tall people below floor level.

Sitting adjustments — For sitting work, adjusting the chair height can often suffice as a way to achieve appropriate heights. However, it is important to make sure there are sufficiently adjustable and sturdy footrests in this case. There are two reasons for doing so: (a) taller stools can be difficult to access and thus sturdy footrests are needed as a type of step, and (b) when a chair is raised to a high position, footrests are crucial to prevent pressure points from occurring behind the knees (see the later principle on pressure points). Note that it is best to mount the footrests on the machine or workbench itself and not rely solely on any foot rings on a chair.

Tilt the work surface — Tilting a work surface sometimes enables working at elbow height, while simultaneously making it easier to see. Drafting tables are a prime example of this concept.

Use tool extenders — The floor is an extremely awkward height from which to work. Long-handled tools are a common way to solve this problem. In some cases, it is possible to use tool extenders.

Equalize height relationships — All of the preceding examples have to do with the relationship between the person and the task, and, as mentioned earlier, are based on the principles of working in neutral postures and reducing exertion. Another category of height relationships, however, is within the equipment itself. The reasons for eliminating these mismatches in heights are to reduce unnecessary motions and needless lifting.

An example of this mismatch is when a product must be lifted over a lip or from one level to another. In these situations adjusting the heights can make the work a lot easier by putting locations in closer proximity to each other, even to the point where items can be slid rather than lifted.

Case Example

 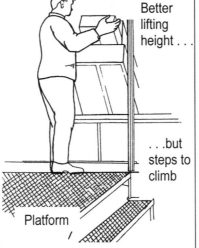 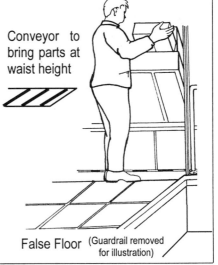

Better lifting height . . .

. . .but steps to climb

Conveyor to bring parts at waist height

Floor

Platform

False Floor (Guardrail removed for illustration)

A good way to supply parts to assemblers is to use chutes and hoppers (see previous principle). However, this can create a height problem in supplying parts to the hoppers.

A good improvement is to build platforms to raise up the parts suppliers. However, this creates steps that need to be climbed regularly.

A better improvement is to install raised floors and then convey boxes of parts in at these elevations. Thus, both the height and the step problems are eliminated.

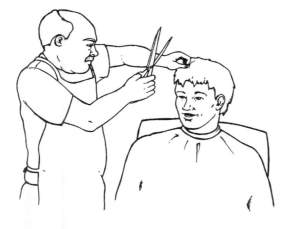

Adjustable height is not necessarily a new concept. In fact, it has been a tradition for barbers and hairdressers. This standard adjustable-design barber chair probably stems from quality and safety issues involved with straight blade shaving. A barber working at an awkward height can easily become more fatigued, resulting in less control over motions and increasing the probability of "errors". With a razor blade at the customer's neck, an "error" of this type could be fatal. Thus, for quality control and meeting customer requirements and expectations, barbers learned the importance of optimizing heights some hundred years ago.

Principle 5

Reduce Excessive Motions

The number of motions required to do a task can have a profound impact on both productivity and wear and tear of the body. Excessive motions can create injury to sensitive tissue and joints, as well as contribute to inefficient use of time. Whenever feasible, motions — in particular, repetitive motions — should be reduced.

In many ways repetitive motions are time wasters, or at least a red flag for such. This does not mean that all tasks involving repetitive motions can be improved (altho many can); rather, these motions often provide little or no value added to the product.

Many of these ideas used to reduce repetitive motions amount to old-fashioned methods engineering, ideas that have perhaps been neglected in a era of high technology. Striving for motion efficiency is a concept that can be readily applied in many workplace ergonomics activities.

Let the tool do the work — One of the best ways to reduce repetitions is to allow machines and tools to do the motions. Machines are good at performing repetitive tasks endlessly, so they should be exploited. Good examples are using a cordless screwdriver instead of a manual one, a nail gun instead of a hammer, or an electric drill versus a hand drill.

There is nothing particularly remarkable about this concept; new applications for power tools are an everyday occurrence. It is worth mentioning, however, because there is often confusion about how to reduce repetitive motions, and sometimes the obvious gets overlooked.

There are countless ways to mechanize repetitive manual motions, including:
- Parts ejectors to remove parts from machines.
- Magazines to feed parts into machines.
- Label dispensers to eliminate the need to pick up a roll of labels and fumble to peel off one.
- Flipper devices.
- Fixtures of multiple tools.
- Total automation and robots.

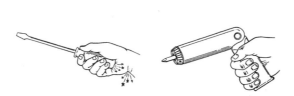

Let tools and machines do the work.

Design for manufacture — Many of the basic concepts of designing for manufacture are principles of ergonomics. Ways to reduce motions include:
- Use snap-in parts, not ones that need fastening.
- Eliminate tangles and "sticky," "springy" parts.
- Eliminate parts altogether.

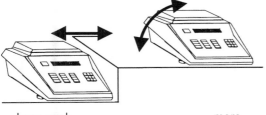

Less work . . .   . . . more

Design for motion efficiency — Another way to reduce motions is to use the most efficient workstation layout possible. Workstation changes to improve heights, reaches, location, and orientation of materials can eliminate many unnecessary hand and arm motions.

For example, in a shipping department, when a mail scale is placed on a level that puts the top of the scale at the same height as the remainder of the worksurface, the packages can be slid on and off, rather than lifted. Similar designs can be used in many assembly operations. Such small changes can have an important impact when the volume and weight of materials are sufficiently high.

Good organization can eliminate some motions altogether. Rather than moving items back and forth, simply arranging them in the proper sequence can eliminate wasted effort. As elementary as this concept sounds, it is surprising how often it is forgotten in everyday operations at work.

Reduce the range of the motion — There is a distinction between a small, insignificant motion and a large sweeping one. Thus, even if a motion cannot be eliminated altogether, it might be shortened. Instances where this rule of thumb may apply include:

- Power tool triggers

- Keyboard keys

- Drill press arms

Slide rather than pick and place — It is almost always better to slide items that must be handled repetitively, rather than pick them up one at a time to place in their locations. Although motions are still required to slide items, the total number is usually reduced. Techniques that enable sliding include:

- Move equipment closer together, equalize heights, and tilt boxes and containers.

- Cut holes in workbenches to permit items or scrap to drop directly into chutes, conveyors, or containers, thus eliminating the need to pick them up and place them repetitively.

Sliding is better than picking up and placing each item.

Rack and pinion — one motion yields multiple turns.

Box stands can be hinged so that the boxes can slide away automatically, eliminating the need to repetitively pick them up and carry.

Motion-saving mechanisms — Similarly, there are numerous mechanical devices that can be brought into play in the effort to reduce repetitive motions:

- Gearing — one turn yields multiple turns.

- Rack and pinion mechanism — one motion yields multiple turns.

- Old-fashion sewing machine pedal — one stroke yields multiple reciprocal motions.

- Yankee screwdriver mechanism — one push yields multiple turns.

- Ratchet — eliminates the motions of releasing hold of a handle only to grasp it again.

Hoppers — Instead of using manual scoops to handle powdered materials, hoppers reduce motions and save time. Similarly, bulk handling systems and pumps can save motions in the handling of both powders and liquids.

Packaging stands that tip — Stands used to hold boxes for packaging of products can be fitted with hinges to allow the supporting surface to drop to an angle that permits the boxes to slide to a takeaway conveyor. Tilters of this type serve to eliminate the motions that are otherwise required for the box to be picked up, carried, and placed elsewhere.

If necessary, rollers can be added to the supporting surface to facilitate the sliding. Furthermore, cylinders can be added to these tilters to activate them with a push of a button. This technique can be used both for stands that are horizontal as well as stands that have already been angled to reduce reaches.

Keep materials oriented — Feeding parts and materials in the correct orientation to a workstation can reduce motions. In particular, a frequent and senseless source of repetitive motions is the need to reorient materials that were once oriented. Ideally, *parts should never be allowed to become jumbled* because it then takes extra work and more motions to straighten them out again. Typically, careful design of chutes and conveyors can keep the materials in their orientation.

Similar examples can be found in how parts are placed on conveyors or on carts to be transferred to the next operation. With some thought, wasted motions can often be eliminated.

Pneumatic tubes used to feed screws to a screw gun provide another example. These feeders keep parts in the same orientation, and they also eliminate altogether the need to pick up and place the screws manually, thus saving many motions.

Automated operations provide helpful insights on methods to keep parts oriented. Because the machines cannot reach into a box of jumbled parts, select one, and place it in the correct position, techniques have been developed to keep parts oriented perfectly. The most common example is the vibrator bowl. These same techniques can also be used to eliminate motions in manual work.

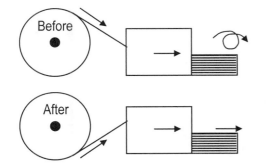

Feeding rolls of paper into the sheeter in a reversed orientation saved multiple flipping motions.

A case example of orienting materials properly to save motions comes from a paper manufacturing operation. An ergonomics evaluation revealed that the first step of a number of operations was to repetitively flip heavy reams of paper merely to put the right side up. Then the ergonomics team discovered that if rolls of paper were fed into the sheeters (the machines that cut the rolls into flat stacks of paper) with the paper coming off the top rather than the bottom, it eliminated the need to flip them afterward — a cost-free improvement.

Mount tools — The classic example of mounting a tool to eliminate unnecessary motions is the bar code scanner. Rather than picking up and then putting down for each use, the scanner can be mounted and the objects with the bar codes slid under. The standard practice of hanging tools from counterbalances also demonstrates good use of this rule. Depending upon the situation, mounting can save time and eliminate a sizable number of motions.

Overhead hook conveyors — A widespread source of repetitive motions is hanging and unhanging parts onto overhead hook conveyors, such as for transporting parts thru paint or wash areas. Much of this is difficult, if not impossible, to resolve, with the exception of certain types of parts that can be dropped: It is possible to automate the release of parts by using scissors hooks or other types of releasing hooks. One type of "automatic" system amounts to having a sim-

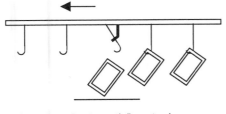

Low-cost "automatic" parts dropper.

ple bar at the dropping station that lifts up each hook as it comes by.

Keyboards — The archetypal instance of repetitive motions in our society is working with the computer keyboard. Design techniques that have reduced motions include:

Improved technology reduces keystrokes.

- Icons, menus, keyboard shortcuts, and mouse devices that have reduced the amount of hand-typed codes that were needed in the past.

- Shortened strokes of keys (especially in contrast to manual typewriters).

- New input devices like touchpads even further reduce the motions needed.

- Voice activation software can reduce the amount of keying, and in certain circumstances can eliminate all or nearly all motions.

Improve work technique — It is not uncommon to see two employees working side by side on the same task, with one employee working smoothly and the other with hectic, exaggerated, and wasted motions. It is important to help employees learn to use the most efficient, least injurious methods. This may involve breaking old habits, but could also be caused by simply not knowing better ways.

Video can provide a powerful training tool.

Videotaping different employees to show both the good and bad methods can sometimes be helpful. Obviously, this must be done with respect and concern for the people involved to avoid any unintended embarrassment or the appearance of being a reprimand. Under the right circumstances, videotapes can be a powerful tool.

Be Creative

One of the barriers to progress is that people have preconceptions about how various tools, tasks, and pieces of equipment ought to look. Fortunately, ergonomics programs in industry can help break these barriers by challenging conventional thinking. With some creativity, hundreds of ways can be found to reduce unnecessary motions, thus improving human well being as well as increasing productivity.

Principle 6

Minimize Fatigue and Static Load

Overloading physical and mental capabilities can contribute to accidents, poor quality, lost productivity, and wear and tear–type injuries. Fatigued muscles are more prone than otherwise to injury, whether acute or cumulative.

Efficiency experts during the early part of the twentieth century placed considerable emphasis on techniques to prevent fatigue, but the topic has been neglected in recent decades. Now with the growing interest in ergonomics, fatigue has regained attention.

In this context, four major causes of fatigue include:

1. <u>Mental overload</u>, which has been well studied within cognitive ergonomics, but exceeds the scope of this book.

2. <u>Work organization</u>, which includes issues like overtime and shift work, is important and well researched, but again beyond the scope of this book.

3. <u>Metabolic load</u>, the concern for which is heavy, exhausting work, has been decreasing thru past decades because of mechanization and will only briefly be reviewed.

4. <u>Static load</u>, which is continuous tensing of a muscle group, is a widespread problem in industry and the focus of this section.

Metabolic Load

Fatigue is an obvious result of heavy, exhausting work, when people sweat and burn calories. This activity is known as high *metabolic load* (as in high *metabolism*), resulting in whole body fatigue. Such heavy work has diminished considerably in the past century, but is still an issue in some workplaces. High metabolic load can be reduced in several ways, most of which are fairly self-evident:

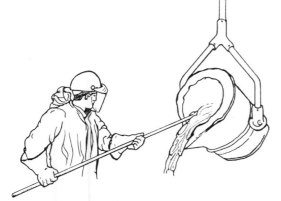

Hot, heavy, exhausting work involves high metabolic load.

- Use machines and powered equipment.
- Spread peak loads over more time.
- Rotate with less-demanding tasks.
- Provide frequent, short rest breaks.
- Add staffing.
- Address other ergonomic issues — force, motions, awkward postures, and environmental conditions such as heat and humidity.

Writer's cramp is an example of localized muscle fatigue caused by static load. Pencil grips help by reducing force.

Products can be presented to the employee to permit working in neutral postures, thereby reducing static load. This example is from an auto assembly plant, where the cars are flipped on their sides to permit work on the underbody.

Static Load

Holding the same position for a period of time, known as *static load*, can cause pain and fatigue. Static load is especially stressful in combination with high force and awkward posture, but the primary concern is the duration of *time* that the muscles are contracted.

One obviously does not sweat and labor under these conditions, but this local muscle fatigue is nonetheless evident; pain and discomfort from static load can be noticeable within minutes. There are additional concerns, however. Continuous static loading of the muscles over months and years can contribute to cumulative trauma disorders. Tensed muscles more or less act as a tourniquet on blood vessels and capillaries, and the reduced supply of oxygen and nutrients is related to tissue damage.

In contrast with metabolic load, problems with static load are much more common by far in modern industry. Furthermore, preventative measures are not altogether intuitive.

General Rules

Reduce force and duration — A good example of static load that everyone has experienced is writer's cramp. After holding onto a pencil for a while the muscles of the hand tire and begin to hurt. It is not necessary to hold the pencil tightly for the discomfort to happen — merely holding it loosely for a long time can cause the hand pain and discomfort.

Prevention of writer's cramp includes: (1) stopping occasionally to stretch, and (2) using a pencil grip, which makes it easier to hold by reducing slipperiness and increasing size of the pencil's diameter.

Work in neutral postures — It is often possible to change the orientation of a piece of equipment being worked on so that it is presented to the person in a way that permits neutral postures. An example in the auto industry is tilting cars so that the underbody work can be done without holding the arms overhead for long periods.

Offload the muscles — The preceding rules help, but to eliminate static load, ways must be found to release the muscles totally. The following provide ad-

Tools held for long periods can fitted with straps to offload the muscles.

The "ring knife" provides an old-fashioned example of how a small tool can be attached to the finger.

Self-closing tool mounted on a fixture.

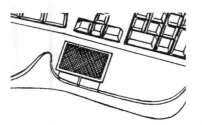

Touchpads eliminate the need to hold the mouse in a static grip.

ditional rules of thumb and examples of how to off-load affected muscles.

Grasping and Pinching Loads

Use fixtures — It is crucial to use fixtures, hooks, or some other way to prevent static grasping of items, whether a part, a tool, or both, if possible. *The hand should not be used as a fixture.*

Use straps and rings — It is becoming increasingly common to use straps to "fixture" a handtool to the hand. Small camcorders offer one of the most common examples of a strap holder. With this arrangement, it is possible to hold the tool in place without needing to constantly employ hand and arm muscles.

Similarly, finger rings can be used to attach small tools to the hand. "Ring knives" have been used for decades in butcher shops and old-fashioned general stores to cut string, providing guidance for other light tools used in the assembly of small products.

Use self-closing tools — Tweezers and clamping tools are available with a design that functions by squeezing to open up the tool, then letting go to hold it in place. This eliminates the need to grasp continually to hold the item with the tool. Locking pliers provide a different version of the same concept, as does a locking trigger on a power tool.

Use alternate input devices — Holding onto a computer mouse for long periods is part of the cause of wrist problems among computer users. Even though the grasping force may be very light, the static load on the hand can be a problem.

A good alternative is the touchpad, which does not need to be grasped at all. Additionally, the space it requires is much less than for a mouse, so the touchpad can much more easily be located within the proper reach envelope.

However, even with the touchpad the whole hand is in a static position. Heavy computer uses may need to switch between various input devices, plus alternate between hands.

Use arm rests — Armrests eliminate static loads on the shoulder in tasks that require outstretched arms. Newer styles of armrests can be attached to the

Bench-mounted armrests can reduce static load on the shoulder.

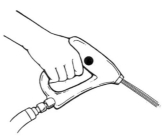

Trigger locks on power tools reduce static gripping.

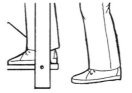

Footrests reduce static load on leg muscles.

workbench or machine. Good, relatively inexpensive adjustable arm rests for chairs have also come onto the market and, in general, should be used for office work.

Use latches — A latch can temporarily keep a lid or door open or hold an object down. Trigger locks on power tools provide a common example. Self-closing door cylinders can serve as a slightly more sophisticated version of the same idea.

Provide load supports — Imaginative use of skids, shelves, and lips on which it is possible to rest a load, even if for only a few seconds, can reduce fatigue. Similarly, slings and chains can help when a load needs to be supported in a certain position for a short period. This concept can work for loading parts into machines, or even heavy fixtures and chucks when they need to be changed.

Use guides and funnels — When a part needs to be inserted into a hole, slot, or fixture, it can be helpful to add a guide or funnel device to help orient parts. This technique can work for both heavy loads as well as small ones to shorten the time muscles are tensed, providing a payoff both for reduced fatigue and greater productivity.

Continuous Standing

The fatigue related to continuous standing is created by static load on leg muscles.

Provide mobility — It is difficult to stand rigid for even a few minutes because of the overuse of the same muscles. In contrast, it is possible to walk for hours, even tho one's overall posture is still standing upright. The difference is that while walking, there is constant shifting in the use of the various muscles in the legs, giving them rest, even if only for split seconds.

Use footrests — For standing jobs, having a footrest available provides a chance to alternate postures from time to time.

Use lean stands — Lean stands provide the advantages of relieving the static load on leg muscles from time to time. Unlike chairs, however, it is possible to revert instantaneously to a standing position for immediate attention to a machine or other work process. Furthermore, unlike chairs or stools, which can take some effort to get into and out of, lean stands are easy to use. Note that the intent is not to remain on a lean stand for long periods; it is simply to obtain some occasional relief from constant standing.

Evaluate feasibility for sitting — There are some jobs where people stand simply because of tradition. To be sure, there are tasks that can only be done from a standing position. However, there others that can, in fact, be done while sitting (or better yet, alternating between sitting and standing). It can be worth the effort to evaluate the feasibility for allowing people to sit.

Lean stands also reduce static load on legs.

Continuous Sitting

It is not always self-evident that sitting for a long period of time can be tiring, since people routinely sit to avoid the fatigue caused by standing. Nonetheless, too much sitting can, in fact, cause fatigue, as anyone who reflects on a prolonged car ride can understand. Cruise control is an excellent example of ergonomics: It is a technique that enables changes in posture and thus reduces fatigue. Sitting in an extremely cramped area, such as an airplane seat or constrained work area, worsens the situation considerably.

Staying *in the same posture* is fatiguing. Ergonomic design provides ways to alternate between sitting and standing so that work is not interrupted, yet postures change as needed.

Provide Movement

Staying in the same position too long can cause fatigue.

The ability to stretch and change postures is important in all of the preceding situations. The human body is designed to move and the emphasis on eliminating motions and using neutral postures for sustained work should not be confused with not moving at all. This topic is addressed in greater detail with Principle 8 — Move, Exercise, and Stretch.

Principle 7

Minimize Pressure Points

Direct pressure or "contact stress" is a common issue in many workstations. In addition to being uncomfortable and an interference with the ability to work, it can inhibit nerve function and blood flow to the point of causing permanent injury. For example, the hand is particularly sensitive because there are (a) large number of nerves throughout the hand and fingers that are typical points of contact, and (b) blood vessels in the fleshy part of the palm where handtools normally press.

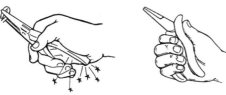

Contoured shapes and padding are two strategies for relieving pressure points.

General Rules

Avoid contact between the body and the tool or piece of equipment whenever possible.

Contour the item to fit the shape of the body at the point of contact.

Provide padding to soften the pressure.

Distribute the pressure over a larger surface area of the body.

Hands and Arms

Provide padding for hand grips — An example of pressure points that almost everyone has experienced is gripping hard onto a pair of pliers or any similar tool. The edges of the metal grips dig into the skin and can create considerable pain and discomfort. Changing the shape, contour, size, and covering of tool handles can more evenly distribute the pressure over more of the palm.

Do not use the hand as a hammer — Depending upon circumstances, the common practice of using the hand as a hammer can cause injury. Squeezing tools and real hammers should be used. In some cases, it may be possible to use a specially padded glove.

Use whole hand loops — Rather than use finger loops for tools such as scissors, providing whole hand loops for heavily used tools eliminates rubbing on the fingers.

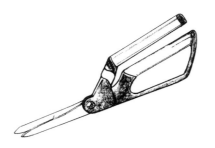

Full hand loops help prevent pressure points.

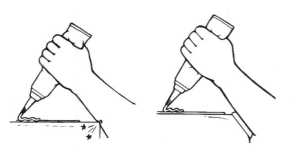

Forearm padding is a common need.

Provide padding for forearms — A similar problem is leaning the forearms against sharp or hard edges for support. Steps for improvement include:

- Pad the edge.
- Round the edge.
- Provide arm rests.
- Redesign the task.
- Change layout to avoid leaning.

Try duct tape and pipe wrap — There are many day-to-day items that can go a long way in reducing pressure points for the hands and arms, at least as a quick fix, including duct tape, pipe wrap, sponges, foam, shop rags, and cardboard.

Legs and Feet

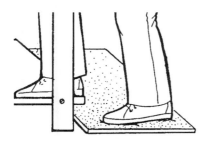

Anti fatigue mat and flat footrest.

Provide floor cushioning — Standing for long periods of time on hard surfaces (especially concrete floors) can damage tissue in the heels, contribute to other leg disorders, and increase fatigue. Options include:

- **Anti fatigue mats,** the usual choice in production facilities where employees stand at a single workstation. A variety of types are available for an assortment of conditions, ranging from oily areas in machine shops to clean room conditions in pharmaceutical labs.

- **Fiberglass grating** can be used in chemical, food, and meat and poultry processing plants where mats are difficult to use. The grating is usually mounted on rubber risers. However, in some cases, they can be placed on metal frames, in which case it is important not to brace the frames too excessively because the whole point is to provide some give.

- **Viscoelastic material** poured into insets in flooring to provide cushioning in oily or other messy environments. The material can be mopped up and, since it is recessed to be level with the floor, involves no tripping hazard, additional congestion, or difficulties in cleaning a mat.

- **Cushioned insoles or heel cups** for mobile staff such as maintenance, engineers, and supervisors,

Grating can provide good floor cushioning in special environments.

where mats are not feasible. Viscoelastic, shock-absorbing materials typically work best.

- **Wood flooring** is more forgiving than concrete, but it usually still requires additional cushioning, such as mats or insoles. Padded carpeting is usually sufficient in offices and similar locations.

Use flat footrests — Foot rings and rails are common on stools and workstations, and they are better than nothing. However, the narrow dimension can create a pressure point on the bottom of the foot. It is better where possible to use a flat surface.

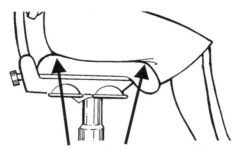

Flat supports are better than rails or bars.

Cushion equipment edges — It is not uncommon for workers to need to lean continuously against hard equipment edges, such as conveyors or workbenches. The first step, of course, is to determine if it is possible to redesign the area or task to eliminate completely the need to lean. If this is not possible, however, then padding can help.

Seating

Contact stress from sitting on hard surfaces is the epitome of discomfort from pressure points, lessened with cushioning and contouring. In recent years the techniques of tilting the seat and having a "waterfall" edge have proved useful in reducing the pressure on sensitive areas behind the knee.

Minimize contact stress by adjusting height.

Proper seat height greatly affects pressure points on the legs. If the seat is too high and the legs dangle, the pressure behind the knee can be excessive. If the seat is too low, the weight of the body concentrates on the buttocks, again creating discomfort. In many ways, proper seat height is determined by finding the balance between the two extremes.

Quick Fixes

A good activity for a workplace ergonomics team is to perform periodic surveys of the facility focusing just on the issue of pressure points. Anti-fatigue mats wear out and other cushions are commonly removed for one reason or another. This type of "dedicated" survey can identify where quick, inexpensive improvements can be made.

Principle 8

Provide Clearance

It is important to have both adequate workspace and easy access to everything that is needed, with no barriers in the way. A common problem in the industrial workplace is insufficient space for the knees, altho every part of the body can be affected — head, torso, feet, and hands.

Problems with clearance are usually readily apparent, altho remedies are not always feasible or easily implemented.

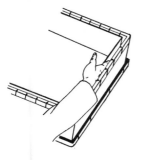

Make sure there is sufficient clearance when ordering equipment.

Design for tall people — In general, the goal is to make sure that tall people have enough clearance, that is, room for the head, knees, elbows, and feet. If tall people can fit, then so can everyone else. To improve access:

- Reorganize equipment, shelves, etc.
- Increase the size of openings.
- Eliminate obstructions between the person and the items needed to accomplish the task.

Provide kneespace — A common problem is lack of knee or thigh clearance on desks, workbenches, or other types of equipment where people sit. As a consequence, people resort to working in contorted, uncomfortable postures that create fatigue and inhibit productivity. A prime example is the situation in which computers are used on traditional desks with file drawers and thick middle drawers under the desk.

Improvements include:

- Thin surfaces, with no hindering drawers.
- Removal of obstacles.

Provide hand clearance — Equally important is having sufficient space for the hands in order to avoid "knucklebuster" injuries and simply to get the job done effectively. Often, these types of problems are not difficult at all to address during design stages or when ordering a certain size piece of equipment. After installation, however, improvements can be difficult.

Use conveyor gates — A specific need in many manufacturing facilities is to have more conveyors or skid bars in order to reduce the amount of heavy lifting. Unfortunately, this type of equipment also tends to block access to work areas.

There is consequently a parallel need to add gates to conveyors or otherwise provide easy access. Approaches include:

- "Drawbridge" gates, which usually require some type of counterbalance, either with a spring or a large counterweight.

- Swing-out gates, which typically are mounted on a wheel and pulled to the side when access is needed.

- Simple hinges for skid bars or sometimes small conveyors.

Design for assembly — Products are sometimes designed so that it is difficult to access areas in order to insert parts or tighten screws. Good planning at design stages can help provide clearance and prevent such problems

Maintainability — Probably the single biggest problem that maintenance personnel encounter in their tasks is lack of clearance. Many activities would be simple to perform, if they could only reach an item and work on it with easy access. Unfortunately, too often the items to be fixed are buried within machines. The remedy is designing equipment with access in mind:

- Removable panels.
- Quick disconnects.
- Relocate frequently accessed equipment.
- Improve configuration.

Provide visual access — A final issue is visual access. Visual access is the ability to see what you are doing or to see dials and displays. As with the other issues noted previously, there are short-term ways as well as long-term ways to remove barriers and change layouts to provide better line of sight.

A common issue is inability to see when moving a cart or lift truck. Equally common are machines where various gauges are distant from the operator's position. General workstations can also suffer from the same problem.

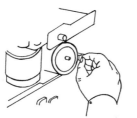

Lack of finger clearance makes assembly more difficult.

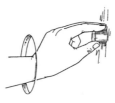

Clearance is needed for maintenance personnel to access equipment that needs repair or replacement.

Visual clearance is important for both safety and production reasons.

Principle 9

Move, Exercise, and Stretch

The human body needs to be exercised and stretched. One should not conclude from reviewing the preceding information regarding the need to reduce repetition, force, and awkward postures that people are best off lying around and pushing buttons.

To be healthy, the human body needs activity. Joints need to be stretched to the full range of motion throughout the day. The heart rate needs to rise for a period of time every day. Muscles need to be loaded on occasion. Unfortunately, most jobs do not promote these activities, and where there is movement or exertion, it is often too much of the wrong sort.

Promoting fitness may increasingly be a part of the workplace.

Warm ups can help prevent injuries and stretch breaks can relieve fatigue from being in a static posture.

Promote physical conditioning

Staying in shape is important; understanding the role of work in promoting conditioning is a key subject. It is an open question whether work ever provided health-promoting conditions, but the twenty-first century may provide new opportunities.

Some suggestions have been made to deliberately design workstations, for example, to require reaches in order to ensure movement and stretching. However, it is probably better to get proper stretching exercise in a fitness center or appropriate sport where the amount and type of movements are better controlled. Furthermore, it would not be appropriate to require a motion that some people may not be able to perform.

An increasing number of companies are in fact providing on-site fitness centers to help promote general conditioning. Clearly, people should use these centers where they exist; however, it would appear to be a difficult challenge to persuade every individual to use them as intended.

Workplace Actions

Activities that can be designed into work include actions such as the following.

Provide warm ups — People who perform heavy tasks should warm up beforehand. Experience in sports has shown the value of warming up to prevent injuries.

Alternate between sitting and standing.

Adjustable-height worksurfaces can facilitate effective work.

There is no one posture that is correct for an eight-hour day. Change positions often.

Provide "Energy Breaks" — People doing sedentary tasks should stop and stretch from time to time. Aerobic activity can also provide benefit and reduce fatigue.

Allow for alternate postures — There is no one "correct" posture that is best for an entire workday. It is important to be able to change and move.

As mentioned previously, one intuitive way to understand the need for variation in posture is to think about a long car trip. By the sixth or seventh hour of driving, drivers are often in discomfort, are fatigued, and resort to sitting in very contorted postures merely to get relief from being in the same position.

Design for sit-stand — An increasingly common way to achieve this movement is with the "sit-stand" workstation. The most basic approach is to design the workstation for a standing posture, then use a tall stool to sit on as needed. One stands until getting fatigued from standing, then sits, and vice versa.

A more sophisticated version of the sit-stand workstation is where the whole worksurface is designed to move up and down at the push of a button. This approach has proved to be of particular value in jobs, such as customer service representatives, that typically involve using a telephone and a computer for the entire day without any alternative tasks to break up the day. Equipment is also available on the market that can simply raise and lower just the monitor and keyboard.

Change chair positioning — For those who sit for long periods, it is important to change the adjustments of the chair:

- Move the seat height up and down a bit throughout the day to get some variation.

- Move the chair back to and fro to alternate between working forward at the desk and leaning backward.

- Shift, move, and change positions often.

Principle 10

Maintain a Comfortable Environment

Humans often do not perform well in less-than-ideal environments. Excessive heat and humidity slow activities; excessive cold hinders effective work. Toxic chemicals can damage health; vibration can injure sensitive tissue.

This principle is more or less a catch-all category in ergonomics. Some issues are normally addressed thru the tools of other fields, such as toxic chemicals and the field of industrial hygiene. Other issues, such as lighting, have gained attention as part of the growth in interest for workplace ergonomics.

Glare is a common problem both in the office and in production areas.

Task lighting often provides a good solution to lighting problems.

Appropriate Lighting

The quantity and quality of light at the workstation can either serve to enhance or obscure the details of the work. Common problems include:

- Glare, whether directly from light sources or indirectly from reflective surfaces.

- Shadows that hide details.

- Poor contrast between one's work and the background.

- Overly bright lighting around computer screens.

A final comment is that poor light can affect more than the eyes. It is not uncommon to contort one's posture to avoid glare or to see better in dim conditions.

Diffuse extremes — Ways to make improvements include providing the following.

- Diffusers or shields to minimize glare.
- Better placement of lights or equipment.
- Indirect lighting to soften bright spots and/or shadows.

Use task lighting — One of the best ways to improve lighting is to use task lights, rather than trying to provide all lighting from the ceiling. There are a number of important advantages to task lighting, including better control of glare, shadows, and brightness at each person's workstation.

Temperature Extremes

Excessive heat or cold while performing tasks can cause discomfort, reduce efficiency, and may contribute to health problems. Typically, the jobs most affected are those where many employees regularly work at temperatures around 40°F or even colder, as in the meat and poultry industry.

In most cases where temperature is an important problem, the source is often inherent in the work and little can be done about root causes. Examples include outdoor work, heat around furnaces, and the cold in meat and poultry plants.

However, some steps can be taken to avoid some specific problems:

Use ventilation deflectors — In cold environments, it is important to insure that no one works directly in the path of blowing air. Diffusers, deflectors, and fabric coverings can be placed over vents.

Monitor heat stress — There is a large body of information regarding the prevention of heat stress, but this is beyond the scope of this book.

Provide shielding for heat sources — *Ambient* heat can be hard to control, but *radiant* heat can be minimized with shielding and clothing of various types.

Vibration

Working with vibrating tools and equipment can potentially cause Carpal Tunnel Syndrome, plus other types of CTDs, such as Reynaud's Phenomenon ("White Finger") as well as fatigue. Whole body vibration can contribute to back injuries and can be experienced from working on vibrating platforms such as found in steel mills and from riding in off-the-road vehicles or trucks on rough surfaces.

Vibration can be reduced thru:

- Better design of tools — such as by improving the internal gearing of tools or by adding vibration-dampening material.

- Isolation mounts — for the vibrating equipment itself or for the walking/working surfaces associated with the equipment.

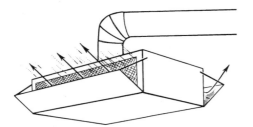

Deflectors can keep cold air from blowing directly on people.

Vibration-dampening material can be designed into tools.

• Preventative maintenance — regular maintenance is typically essential in keeping vibration levels low in tools and equipment.

• Dampened tool grips — such as by isolating the grip from the vibrating tool or adding wraps. One caution in adding wraps to a grip is to insure that the circumference does not become so excessive that grip force is increased or the tool becomes awkward to handle.

• Vibration-dampening gloves — which can be used in some circumstances, altho typically only as a last resort. These gloves can be thick, which can reduce grip strength and make hand work unwieldy.

• Deadblow hammers — the heads for which are filled with shot rather than being solid and thus absorb shock.

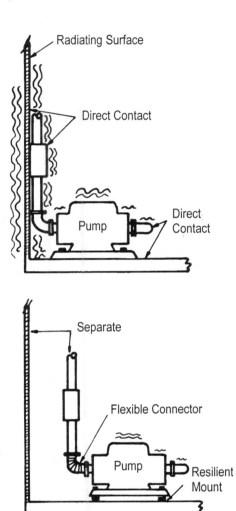

Hundreds of techniques are available to reduce workplace noise.

Noise

Noise has been addressed thru the years as a cause of hearing loss and a contributor to fatigue and stress. However, in some types of workplaces, most typically offices, noise has never been considered a health problem because the levels are usually nowhere close to exceeding the permissible limits.

However, in many of these cases noise is now surfacing as an issue as part of the overall ergonomics effort. Noise can certainly be a nuisance and can mask communications and therefore part of the overall effort to design tasks to fit people.

Color

Colors can affect tasks in a number of important ways:

• Small parts on a traditional white assembly table can be difficult to see. Contrasting worksurface colors can make them visible.

• Matte colors can reduce glare and eyestrain.

• Color on walls and good use of art can enhance a work area. (Tile and mosaics can be used in sterile or clean environments.)

11

Applications

This chapter expands on the basic principles of ergonomics as they are applied in various settings. The examples chosen fall into three general categories:

- Common industrial work settings.
- Computer workstations and chairs.
- Special applications that help demonstrate use of principles.

Please note the following:

1. The illustrations for each application provide only one way of addressing the issues. There may be other approaches for each application that are equally good.

2. Hundreds, if not thousands, of other examples could have been used.

3. The point is to understand how the basic principles of ergonomics can be applied in various settings.

SITTING WORKSTATIONS

Common issues

- Work surfaces not at appropriate heights
- Long reaches for materials
- Work surfaces with hard edges
- Inadequate clearance for thighs and knees
- Inadequate work area to perform all tasks
- Shadows or glare

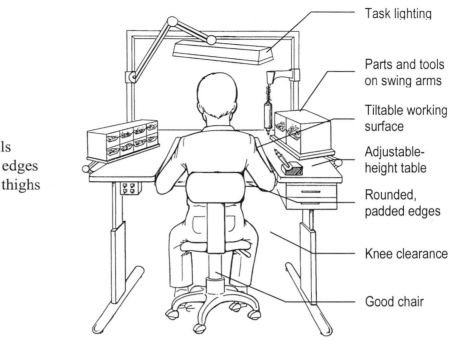

Task lighting

Parts and tools on swing arms

Tiltable working surface

Adjustable-height table

Rounded, padded edges

Knee clearance

Good chair

STANDING TASKS

Common issues

- Static load on legs
- Awkward back posture from too severe a sway in the lower back when standing straight up
- Awkward back posture when bending over to perform tasks
- Pressure points from hard floor

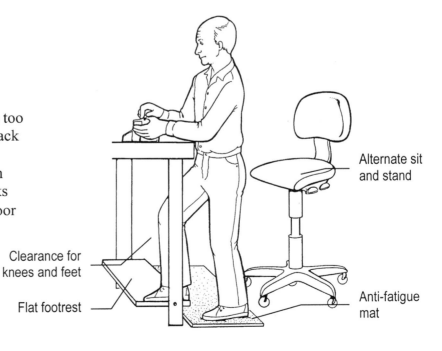

Alternate sit and stand

Anti-fatigue mat

Clearance for knees and feet

Flat footrest

LABORATORY

Adjustable tilt
for microscope

Padded, forearm support

Common issues

- Bent neck and back using microscope
- Pressure point on forearm
- Extremely constrained, static posture
- Eye strain and fatigue

Not shown, but typical problems in labs:
- Lack of knee space
- Awkward heights for various tasks
- Pipetting
- Standing

Options for improvement

- Modular storage cabinets for work areas
- Risers, lab jacks, adjustable height surfaces
- Shorter tubes, electric pipettes
- Fatigue mats and foot rests
- Energy breaks; sit/stand workstations
- Multiple focal length magnifying glasses
- Replace microscope with video camera and monitor (an increasingly feasible alternative that permits movement of the upper body)

MANUFACTURING WORKBENCH

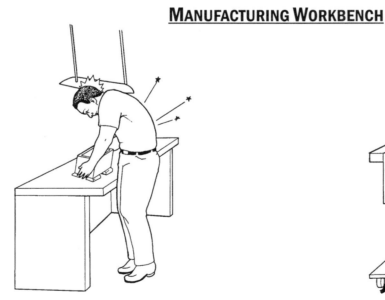

Common issues

- Work height too low (or high)
- Lack of head clearance
- Tools placed haphazardly
- Constant standing
- 6-foot long bench creates congestion

Options for improvement

- Modified die cart, foot pump adjustment
- Light mounted to (and adjusts with) bench
- Tool holder
- Sit/stand capability; built in footrest
- Small workstation on casters creates space

POWER DRIVERS

Issues

- Grip force
- Static grip
- Weight of tool
- Vibration
- Repetitive finger motions to trigger
- Repetitive hand and finger motions to insert fasteners
- Repetitive arm motions to pick up and replace tool
- Shock when fastener bottoms out

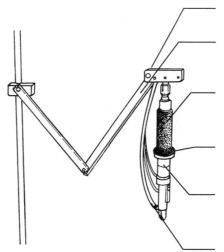

Tool balancer

Torque bar
reduces shock

Padding provides friction
surface to reduce grip force,
plus shock/vibration dampening

Collar reduces
grip force

Sleeve trigger
reduces motions

Tube-fed fasteners
reduces hand and
arm motions

HOSES AND CORDS

Issues

- Heavy cord
- Congestion
- Hand as fixture
- Stiff hoses and couplings

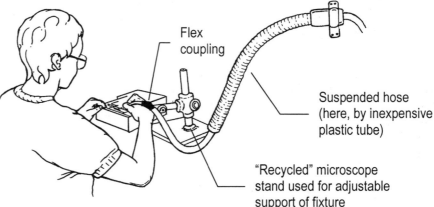

Flex
coupling

Suspended hose
(here, by inexpensive
plastic tube)

"Recycled" microscope
stand used for adjustable
support of fixture

TOTES/BASKETS/PANS

Issues

- Pinch grip
- Sharp edges on grip
- Small grip
- Heavy
- Stick together
- Protruding

Handholds, round
and sufficiently large

Lightweight

Nestle
loosely

Easy to
slide

HAND TOOLS

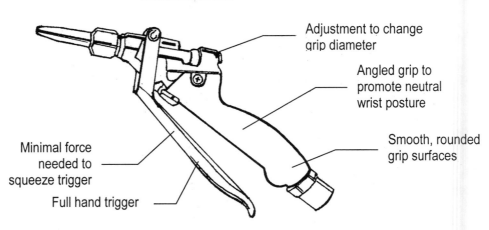

Adjustment to change grip diameter

Angled grip to promote neutral wrist posture

Smooth, rounded grip surfaces

Minimal force needed to squeeze trigger

Full hand trigger

Common issues

- Bent wrists to operate tool
- Hard trigger pull
- Heavy weight of tool
- Slippery grip
- Inappropriate grip diameter

- Static grip to hold tool for long periods
- Static load on shoulder to extend tool
- Pressure points on grip
- Heat or cold
- Vibration or shock as tool deploys

Options for improvement

- Hinged grips   (for wrist posture)
- Detachable, multi-sized grips   (for different hand sizes to reduce force)
- Vacuum triggers   (to reduce force)
- Engagement triggers   (to reduce finger motions and force)
- Counterbalances   (to reduce load on arm)
- Hand straps and clips   (to reduce static load)
- Vibration and shock-absorbing grip wraps

Dedicated Tools

It can often be helpful to have *dedicated* tools or pieces of equipment. Instead of using generic items, it is often better to modify them or have multiple varieties available to fit specific needs.

A good example is the meat industry, where instead of a single type of knife, there are short stubby blades, long thin blades, flexible blades, two hand drawknives, etc. More common is the use of carts — instead of one generic type used thruout a whole facility, it often is useful to have available a large variety of sizes and shapes. Providing both pistol grip and inline

power tools at the same workstation is a third example.

Similarly, instead of having only one tool or piece of equipment available, it is sometimes better to have multiple ones available:

- One at each machine
- One at either side of a large piece of equipment

It is more expensive to buy more items, but the timesavings can cut costs in the long run.

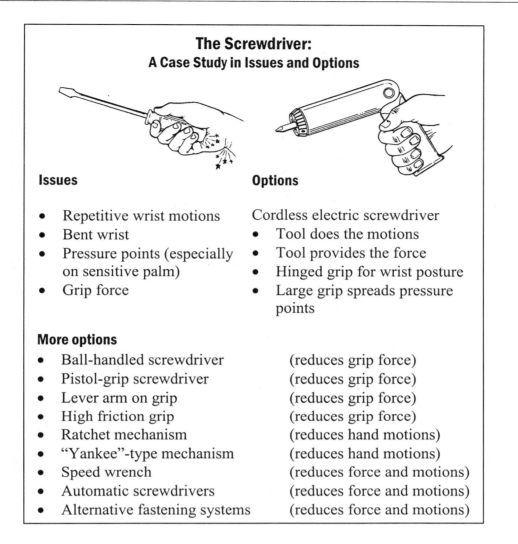

The Screwdriver:
A Case Study in Issues and Options

Issues

- Repetitive wrist motions
- Bent wrist
- Pressure points (especially on sensitive palm)
- Grip force

Options

Cordless electric screwdriver
- Tool does the motions
- Tool provides the force
- Hinged grip for wrist posture
- Large grip spreads pressure points

More options

Ball-handled screwdriver	(reduces grip force)
Pistol-grip screwdriver	(reduces grip force)
Lever arm on grip	(reduces grip force)
High friction grip	(reduces grip force)
Ratchet mechanism	(reduces hand motions)
"Yankee"-type mechanism	(reduces hand motions)
Speed wrench	(reduces force and motions)
Automatic screwdrivers	(reduces force and motions)
Alternative fastening systems	(reduces force and motions)

Triggers

Single finger triggers are traditional and often the easiest type for occasional use. However, if the use is sustained, a single finger can quickly become fatigued and suffer cumulative trauma. Triggers and switches come in various types, including some that are much better for sustained work:

- Full-hand triggers spread the load among all the fingers. If the tool is awkward to hold while using a full-hand trigger, a strap or collar can help secure it to the hand.
- Sleeve triggers or palm triggers eliminate the need for any separate triggering motion of the hand. As the tool is put into place and pushed, it engages.

- Vacuum triggers can be used with certain machines and tools. The "trigger" is a mere hole in the side of the hollowed tool, which is put under vacuum. When the finger is placed over the hole, the seal of the vacuum is completed, triggering the tool with no moving parts in the tool itself and no exertion at all for the finger.
- Photoelectric sensors, or "finger sweeps," as mentioned previously, can be used to replace the traditional palm buttons.
- Proximity switches have many uses, including replacement of palm buttons.
- Knee, elbow, and foot activated switches all have their uses.

WORKSTATION DESIGNED AROUND A FIXTURE

A good concept is to design a workstation around a fixture rather than use a workbench. The following is a case example:

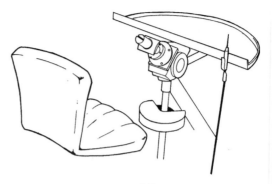

| **Before** | **After** |
| **Traditional Workbench** | **Designed Around a Fixture** |

- Static grip to hold and manipulate parts
- Awkward arm posture
- Repetitive motions
- Inadequate chair (not shown)
- Improper height of work bench
- Long reaches
- Inappropriate tool grip

- Fixture for parts
- Angled fixture to orient part for easy access
- Powered rotation of fixture
- Car seat (with push button controls!)
- No work bench; good seat height adjustment
- Parts and tool racks in reach envelope
- Improved tools (not shown)

SPECIALIZED EQUIPMENT

Commercial lawnmowers

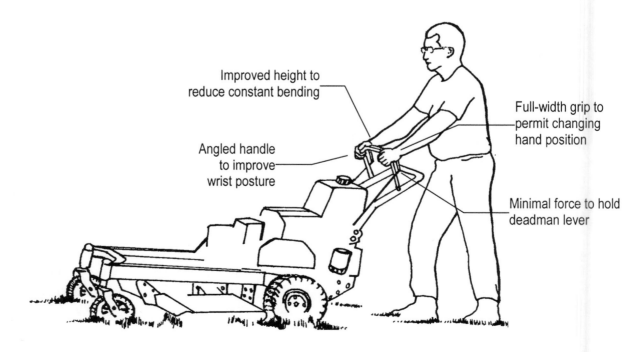

Improved height to reduce constant bending

Angled handle to improve wrist posture

Full-width grip to permit changing hand position

Minimal force to hold deadman lever

PALLETIZING AND MATERIAL HANDLING

Common ergonomics issues

Repetitive lifting, bending, twisting, and reaching. Heavy loads.

Options for improvement

Scissors Lifts

- Powered lift tables
- Spring-loaded lift tables
- Zero-clearance lift tables
- Lift tables with Lazy Susans
- Lift tables with roller conveyors
- Lift tables that also tilt
- Automatic palletizing systems

Stretch wrappers

- No-bend manual stretch wrappers
- Automatic stretch wrappers

Conveyor systems

- Skids and slides
- Roller conveyors
- Screw (or auger) conveyors
- Ball conveyors
- Flex conveyors
- Powered belt conveyors
- Air tube conveyors

Mechanical assists

- Powered, articulated arms
- Miscellaneous cranes and hoists
- Vacuum hoists

Alternative Methods

- Hoppers
- Chutes
- Guides and funnels
- Dollies and carts
- Gurneys
- Runways
- Pumps
- Overhead monorail systems
- Cell production (to reduce handling)
- Layout optimization (to reduce handling)
- Pressurized air (hovercraft concept)
- Bulk handling
- Lift trucks with custom grabbers
- Pick-and-place robots

Height of palletized materials	Type of pallet stand
Short stack:	Knee-high, fixed height stand can be OK
Medium stack:	Consider a lift table
High stack:	Consider a lift table recessed into the floor

Note: This simple guide obviously depends on frequency of lifts, type of material, and other factors. It also assumes multiple layers and not single tall items.

WAREHOUSING

Common Issues

Manually picking stock or orders can require repetitive bending and lifting of often-heavy materials, usually in a constrained area due to the racks. Unfortunately, there does not appear to be a feasible design that addresses the core of the problem. In some circumstances, however, certain techniques have provided some gains:

Options for Improvement

- Order carts with adjustable heights
- Lazy Susans for especially heavy products
- Lift tables for especially heavy products
- Raise bottom rack off floor about one foot
- Use fixed height pallet stands for short stacks of product
- Store large, awkward products separate from racks

Ideally, product should be stored above the knees and below the shoulders (in reality this goal is often not feasible because of space constraints).

MACHINING

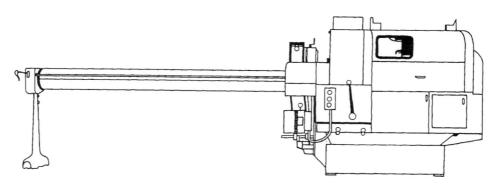

Common Issues

- Heavy lifting and exertion to load stock
- Heavy exertion to change tooling
- Long reaches to access tooling and handle parts
- Long reaches to make adjustments
- Lifting of heavy totes and baskets
- Repetitive hand and arm motions to load and unload parts
- Repetitive hand motions to deburr
- Continuous standing on hard floor

Options for Improvement

- Shuttles to load parts
- Equalize worksurface heights to permit sliding of materials
- Skid bars and guides to support loads while putting into and removing from machine
- Skid bars, conveyors with gates, or flex conveyors to handle totes
- Air or hydraulic cylinders to do heavy pushing
- Lift tables and tilted parts stands
- Lean stands and footrests
- Anti-fatigue mats or cushioned insoles

MAINTENANCE

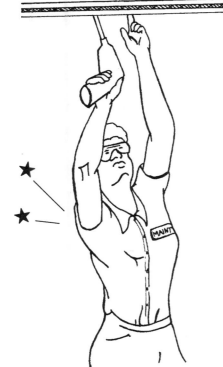

Maintenance tasks are often overlooked for ergonomics evaluations, which is unfortunate given the potential benefits. Good ergonomics of maintenance can eliminate some of the awkward activities associated with maintenance and repair. Moreover, these same considerations can also reduce downtime for repairs, improve inspection of critical components, and reduce failure-related accidents.

It is often difficult to evaluate and improve maintenance tasks because of several important constraints:

1. Tasks are often performed only sporadically, making it difficult to evaluate activities.

2. There is usually no single workstation to evaluate, again making it more difficult to spot issues.

3. Improvements are often achieved only thru design of new equipment or facilities.

Common issues

- hard to get at equipment, resulting in:
  - contorted postures
  - long reaches
  - working at awkward heights
- heavy exertion, often from awkward positions
  - manhandling equipment
  - carrying loads
- repetitive motions (not classical assembly-line work, but still repetitive)
  - significant numbers of motions of hands, arms and trunk in day
  - some repetitive tasks (e.g. screwdriver)
- pressure points
  - hands, arms, and body against tools and equipment
  - standing and walking continuously on hard surfaces

- static load
  - using hands as fixtures
  - holding arms overhead
  - constant standing
- poor access and clearance various handtool issues, including
  - grip design
  - vibration
  - hand exertion
- environmental issues
  - vibrating tools and equipment
  - temperature extremes
  - lighting issues: glare, darkness, shadows
- confusing dials or indicators, or operation of controls

Options for Improvement

The issues are so diverse that it is not possible to provide one set of ideas; rather, all the basic principles and rules should be applied as needed (see especially Principle 8 *Provide Clearance*).

MEATPACKING

The experience in the slaughter industry is important for all ergonomics practitioners to recognize. The following is a list of the types of changes that have led to success in meatpacking.

Ways to reduce exertion

Tumblers — Hams are placed in large tumblers that soften them up to be easier to cut.

Improved coolers — Changes in how the sides of beef or pork are stored in cold meat lockers helped insure that the meat was not frozen when it was cut up.

Fixtures — A variety of innovative fixtures have been developed to reduce strain on the "holding hand." Many of these fixtures are simple hooks.

Mechanical assists — A variety of powered assists have been invented to eliminate high-exertion tasks, such as pulling apart meat from bone, removing the hides, or inserting core samplers into bins of ground meat.

Cart design — Larger wheels, better floor, adding grips to carts, and using various types of powered tuggers have all help reduce the force needed to move carts around.

Ways to improve posture, heights, and reaches

Adjustable tables — To accommodate employees of various heights, work surfaces have been designed to adjust in height, some even at the push of a button.

Adjustable stands — When it was not possible to adjust the work surfaces, the meatpacking companies improved the types of stands that were available to permit height adjustments, some at the push of a button.

Tub tilters — It had been common to work bending over into various types of tubs; now devices that raise up and tilt these tubs are available and are being phased into operations.

Tilted work surfaces — Some conveyor cutting lines have been tilted to enable workers to work in more upright postures.

Height changes — Countless work surfaces have been modified, including lips and barriers between various pieces of equipment that created extra work.

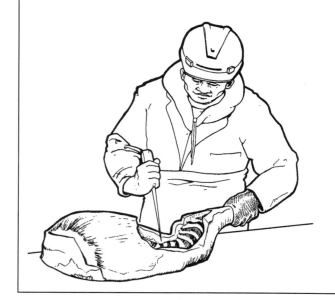

Knife Work

Since cutting knives are so important, significant efforts were devoted to improvement:

- Better control of knife sharpening systems
- Knife grips that are easier to hold
- Better control over cooler temperature to prevent having to cut frozen meat
- Changing heights and tilts of cutting tables to get a better angle of attack
- Using massive tumblers to soften the meat (typically hams) before being cut
- Improved training programs to help workers keep their knives sharp and use good techniques

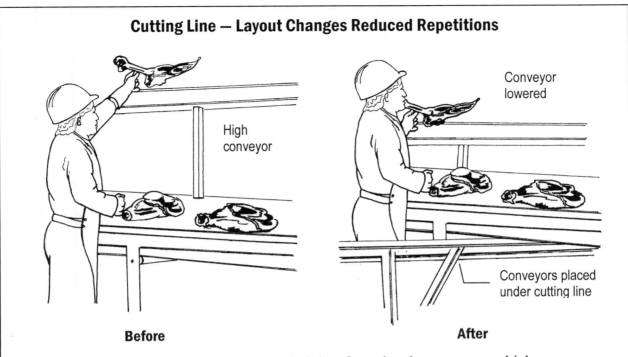

Cutting Line — Layout Changes Reduced Repetitions

Before | **After**

High conveyor

Conveyor lowered

Conveyors placed under cutting line

A common problem was the height of overhead conveyors, which traditionally had been placed quite high, and thus causing unnecessary arm motions to toss the meat or bones upwards. Ways were found to lower the conveyors and even place some under the cutting tables so that upward motions were reduced.

Ways to reduce motions

<u>Sliding</u> — Ways have been found to slide products, rather than lift them.

<u>Motion efficiency</u> — Layout changes have reduced the number of motions required to do certain tasks.

<u>Drop overhead conveyor heights</u> — By lowering overhead conveyors, the need to throw products upward was eliminated and arm repetitions were considerably reduced.

<u>Automation</u> — Some important advances occurred, stimulated by the ergo programs. For example: (a) machine vision-controlled pork belly sorters, and (b) mechanical presses to squeeze meat off beef neck bones.

Standing

<u>Lean stands</u> — Most packers have experimented with providing lean stands on cutting lines (with mixed success).

<u>Cushioned grating</u> — Because it is difficult to use anti-fatigue mats the concrete floors in this environment, where sanitation is such an issue, some packing plants have placed fiberglass grating on mounts about ½ inch off the floor, which then provides sufficient cushion.

<u>Footrests</u> — Various ways of providing something on which to place one foot or the other on from time to time.

THE COMPUTER WORKSTATION

Rules
(Required)

Tips
(Exceptions are possible)

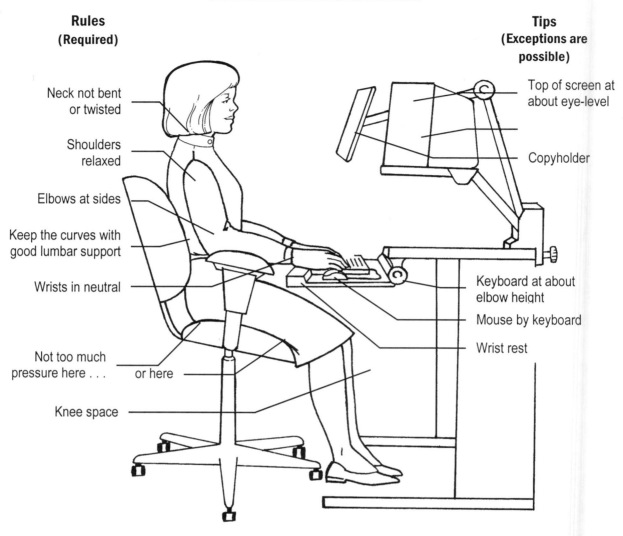

Neck not bent or twisted

Shoulders relaxed

Elbows at sides

Keep the curves with good lumbar support

Wrists in neutral

Not too much pressure here . . .   or here

Knee space

Top of screen at about eye-level

Copyholder

Keyboard at about elbow height

Mouse by keyboard

Wrist rest

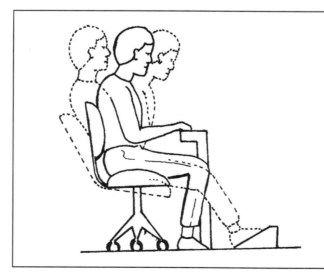

NOTE:
There is no one posture that is "correct" for an eight-hour day — the body needs movement and change. The point is to (a) have a good starting position from which one can vary, and (b) avoid having the furniture force someone into a single, contorted posture.

KEYBOARDS

Contoured
(mounded and angled)

Touchpad,
integrated with keyboard

A variety of modifications in keyboards have been introduced in recent years to address a number of ergonomics issues. While some designs are quite radical looking, others incorporate small but meaningful changes, such as contouring the keyboard. Not all users need anything different than the traditional straight keyboard, but heavy users should consider the newer designs.

Technology for computers and input devices is obviously changing very rapidly. That is why understanding the basic principles of ergonomics is so important, rather than memorizing a series of rules that change depending on the situation. When the next wave of devices arrives it will be possible to make one's own evaluations and adaptations.

Common ergonomics issues
- Bent wrists to align fingers with keys
- Bent wrists to support palms on desk while typing
- Long reach to mouse
- Static load on hand to hold mouse
- Arm and wrist motions to manipulate mouse
- Repetitive finger and hand motions to type

Options for improvement
- Contoured keyboard   (puts wrists in neutral posture)
- Use wrist rests   (puts wrists in neutral posture)
- Tilt keyboard   (puts wrists in neutral posture)
- Place mouse adjacent to keyboard   (reduces reach)
- Use voice activation   (reduces finger and hand motions)
- Use touchpad   (reduces reach, motions, and static load)

CHAIRS

Good chairs with considerable adjustability are needed to (a) accommodate individual sizes and needs, and (b) permit variation of posture when sitting for long periods of time. Also, with a good chair, shortcomings in other equipment can be minimized.

Shop chairs are on the market that have all the necessary ergonomic features, yet are sturdy and have durable coverings suitable for dusty or gritty areas.

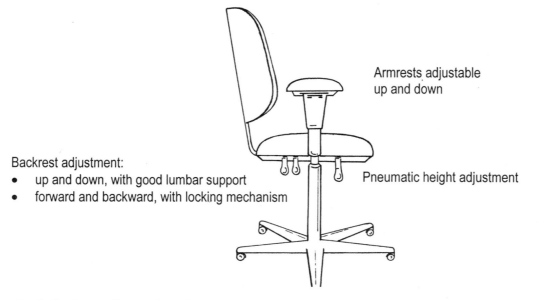

Armrests adjustable
up and down

Backrest adjustment:
- up and down, with good lumbar support
- forward and backward, with locking mechanism

Pneumatic height adjustment

Basic features of a good chair

A general rule of thumb is that persons who routinely sit for more than four hours per day should be provided with chairs that have these basic features:

- The chair must have good lumbar support.

- Seat height adjustment should be easily and immediately accessible. This is typically accomplished with a pneumatic lift operated with a simple lever.

- The chair back should be adjustable separately from the seat. Adjustments should be in two dimensions: (a) height and (b) forward and backward tilt, with a locking mechanism.

- Armrests should be adjustable up and down.

- The front edge of the seat should have a front edge relief (rounded "waterfall" cushion, not square).

Ideal features

Distinction between basic and ideal features is mostly a matter of professional judgment, but it commonly includes:

- seat pan tiltable
- rocking feature with tension adjustment
- armrests adjustable toward and away from the body
- armrests tiltable

Other features

Cushioning — Configuration and thickness of cushioning varies considerably among different manufacturers and chair styles, and is more of a personal choice rather than ergonomics recommendation. For this reason, it is common to make several different styles of chairs available to employees, as long as the core ergonomics requirements are met.

Durability — Durability is obviously important for heavy use. Ease of use of the controls is another factor. Aesthetics, type of fabric, and other issues also affect suitability of a chair, but are beyond the scope of ergonomics.

Size — Many chairs come in various sizes of seat pans and backrests, with selection obviously dependent upon individuals.

A note regarding costs

Chairs do NOT need to be expensive to have all the necessary basic features. For example, the national chains of office furniture and supplies offer good chairs with all the basic features for as little as $130.00. These chairs may or may not have sufficient durability for constant use (although they are probably adequate for intermittent use), but the point is that the adjustability itself does not need to be expensive.

Low-cost options

In some cases, new chairs may not be necessary or feasible. Improvements can be found in the following ways:

- Add support cushions (foam, fabric, bubble wrap, etc.) to armrests to improve comfort or to raise armrest height.

- Add a commercially available lumbar support cushion to the chair back to facilitate the natural curve of the back.

Lower back

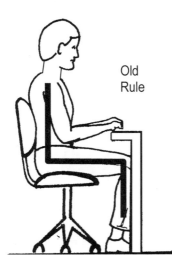

Old
Rule

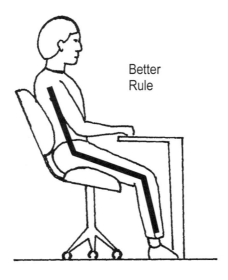

Better
Rule

Yet
Better
Rule

Good lumbar support in a chair is usually the most important consideration. If it is not possible to procure new chairs, lumbar support cushions can help greatly. There are many styles available on the market. At times, even a small pillow or rolled towel will do.

The overall design of the chair is important as well. In the past, many chairs were designed to facilitate the old "90–90–90" rule for the "correct" typing posture. This rule was based on the idea that the body was supposed to fit square corners — 90° at the knees, 90° at the hip, and 90° at the elbows.

It is still acceptable to consider this rule, as long as it is used as a reminder for a starting position from which one can adjust. It is easy to remember 90–90–90 and it can keep people from sitting in extremely distorted postures.

However, there is no real scientific basis for the rule; the human body does not work best at right angles. Studies show that it is better to keep the angle between the thigh and the spine at about 120° and to lean back in the chair at about 120°. (Of course it is still important to (a) keep the S-curve, (b) have good lumbar support, and (c) adjust everything else accordingly.)

The benefits of this sitting position are evident:

- The backrest now supports much of the upper body weight, transferring it from the spine.

- The spinal column tends to stay in its curved alignment because the pelvis is less angled.

But to reemphasize, there is no one single posture that is correct for an eight-hour day. The body needs to change and adjust. Thus, people should change posture from time to time, adjusting the chair as appropriate to keep a more or less neutral posture, yet still getting diversity.

This is one reason why good, adjustable chairs are so important. They can provide the ability to change posture from time to time, yet promote comfortable, neutral postures.

Arms and shoulders

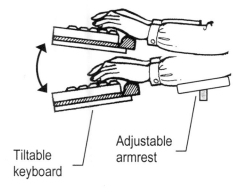

Tiltable keyboard

Adjustable armrest

Holding the arms out in front while typing places a static load on the shoulders. Good arm rests on chairs can eliminate this problem almost entirely. Because the length of peoples' arms varies considerably, having adjustable armrests is crucial.

- If the chair has adjustable armrests, they should be placed at about the same height as the keyboard.

- If the chair has fixed armrests that are too low, a quick fix is to wrap some padding around them.

Some keyboards and keyboard holders can be tilted. Consequently, if the forearms are sloped down, the keyboard should be sloped downward accordingly (a "negative tilt"). If the forearms are sloped upward, such as when leaning back in the chair, the keyboard should be at a "positive tilt."

Note that the elbow does not necessarily need to be at a 90° angle, especially if it is in combination with appropriate equipment, such as a sloped desk or a tiltable keyboard tray. (Maintaining neutral wrists and relaxed shoulders is generally what is most important.)

Poor — desk too high   **Good**

Hunched shoulders

Elbows winged out

Shoulders relaxed & elbows at sides

A static load on the shoulders can be worsened if the keyboard is too high, causing people to compensate by winging the elbows out or hunching the shoulders. Obviously, adjusting the heights of chairs and work surfaces enables working in neutral postures and thus eliminates this source of fatigue and discomfort.

Foot Rings on Stools

If tall stools are used for high workstations, it is usually better to attach the footrest to the workstation legs or frame, rather than to rely strictly on a foot ring mounted on the stool. The foot ring requires keeping the leg continuously in the same position, which can be fatiguing after a time. Moreover, the foot ring can be unstable under some circumstances.

EXAMPLE OF LEISURE-TIME APPLICATION — BICYCLES

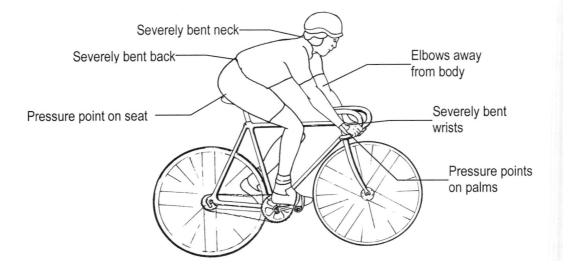

Severely bent neck
Severely bent back
Pressure point on seat
Elbows away from body
Severely bent wrists
Pressure points on palms

Note the number of ergonomics issues found in the design of the traditional road bike. These factors help explain why many avid bicyclists have physical problems and injuries, especially with their wrists and neck.

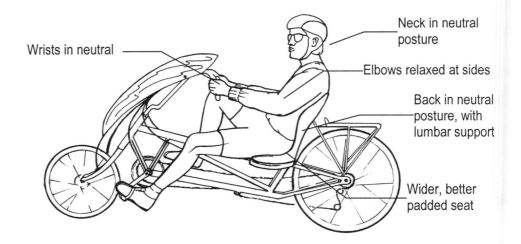

Wrists in neutral
Neck in neutral posture
Elbows relaxed at sides
Back in neutral posture, with lumbar support
Wider, better padded seat

Then, review the ergonomics advantages of the recumbent bike. It is also important to recognize that a recumbent bike holds the world speed record at 65.48 mph. When recumbent bikes first appeared in the early twentieth century, they were faster than the existing bikes, but because of the controversial design were banned from competition. Consequently, the design fell by the wayside for nearly a century; however, it is now enjoying a surge in popularity, primarily because of the increased comfort.

12

Cognitive Ergonomics

What is often thought of as being human error may in fact be poor ergonomics. Many mistakes and errors that people make may really be attributable to poor design, such as confusing dials or controls that do not operate as expected. These are often-overlooked issues that have great implications for process improvement in industry.

The focus of this book is physical ergonomics, but it is helpful to have at least an overview of basic cognitive concepts. This helps round out an understanding of what it means to "fit the task to person" and "make things user-friendly." It is also worth noting that cognitive ergonomics comprises a larger portion of the total field of ergonomics than does physical ergonomics. A larger portion of the profession and research has been devoted to cognitive rather than physical issues.

With careful study of the errors people make, it is possible to predict human reactions, then design tools and systems to take these reactions into account. With good ergonomics, errors can be reduced in products and processes, ranging from simple household appliances, such as a cooking range, to complex control panels, such as for an air traffic control tower.

To Err Is Human

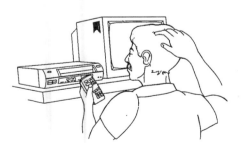

Cognitive errors fall in three categories:

1. <u>Perception Errors</u> — The operator making the error did not grasp the needed information for any number of reasons. For example, the signal or message was not clear, there were other distracting signals, or the person was not trained in the meaning of the information.

2. <u>Decision Errors</u> — The operator did not respond to the signal or information. Perhaps other decisions also had to be made quickly and he or she decided that the signal was not important or a priority. Alternatively, the person may have judged the situation incorrectly.

3. <u>Action Errors</u> — The operator reacted, but activated the wrong control or activated the correct

Operation of single faucets (such as outdoor spigots for garden hoses) have been fairly standardized and follow the popular rule for activation — "righty-tighty, lefty-loosey."

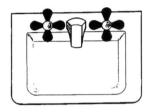

Double faucets on the other hand are not standardized.

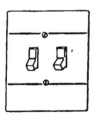

Vertical switches follow a stereotype and cause few errors (assuming one switch per light).

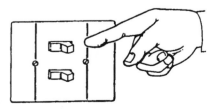

Horizontal switches violate the stereotype, provide no hint of correct operation, and result in confusion and errors.

control improperly. Perhaps the controls were not laid out or did not operate as the person expected. Alternatively, perhaps a control was inadvertently activated even when there was no signal or decision made.

General Rules

Standardize — Many errors are caused by inconsistencies in how information is displayed and how controls work. To prevent mistakes, a general rule is to insure that similar devices work the same way. Agreeing upon a standard helps prevent errors.

In the workplace, lack of standardization between pipe fittings, pressure valves, and other equipment has caused mistakes with serious repercussions. A standard could be agreeing to wire the controls of several pieces of equipment in the same way, so that an operator can easily switch from one to another. A further example is color-coding wires or pipes. A more formalized standard is a government regulation for labeling hazardous chemicals. Definitions are also standards — agreeing to call things in a uniform way to prevent confusion.

It sometimes does not matter which system is adopted, as long as it is always the same. Other times, an arbitrarily set standard may conflict with users' perceptions. To prevent the latter problem, the standard should conform to human perceptions.

Use stereotypes — A stereotype is a commonly held expectation of what people think is supposed to happen when they recognize a signal or activate a control. Good design should take advantage of these perceptions and expectations.

The concept of a stereotype is closely related to that of a standard, but much less consciously determined. Whereas a standard is a formal agreement to eliminate inconsistencies, a stereotype is an informal convention that has evolved through time. A good standard often follows a stereotype. Conversely, a standard that has become culturally ingrained through widespread use can become a stereotype. Examples of stereotypes are:

- Using red to mean stop or danger.
- Flipping switches up turns items on.
- Turning a dial to the right or pushing a lever forward increases speed or power.

Designs like the above (typically candy machines) are often prone to error because the buttons are located some distance from the actual product.

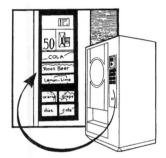

In contrast, this design (typically soft drink machines) often has the symbol of the brand imprinted directly on a large activating button, reducing chance of error.

Gas and water meters with reversing analog dials are complex and error-prone.

Icons can present information effectively.

Link Actions with Perceptions — Ideally, there should be a strong relationship between the perception of the need to take an action and the action itself; that is, a compatibility between a display of information and a control. Good design means configuring items so that it is self-evident what one is supposed to do.

An example is a complex piece of machinery that has a control panel with a multitude of dials. The controls that govern these machine functions should be linked closely with the dials, so that it is intuitive which control affects which function and which dial.

Mnemonics also help in this regard, such as on a car gearshift where the "R" stands for Reverse and the "P" for park. Computer keyboard commands have also exploited this concept. A "CONTROL-P" for printing is much easier to remember and more closely linked to the perception of the need to print than is some arbitrary numeral or other letter.

Another good example of this principle is the use of recorded verbal warnings in commercial airplane cockpits. For example, if an airplane goes too low, a recorded voice comes on commanding urgently, "Pull up! Pull up! Pull up!" This system links very quickly the perception of danger with the action of increasing altitude and is far superior to previous systems of warning lights and buzzers, or of relying on the pilots to check the altimeter from time to time.

Simplify Presentation of Information — Too much information is sometimes provided, or it is provided in too complex a fashion. In general, good designs provide simplified displays.

The traditional gas or water meter dials (that alternate clockwise and counterclockwise directions for each digit) are examples of complex presentation of information that is nearly impossible to read correctly without careful study and double checking. Even the professionals from the utility companies make mistakes. A common good example is the use of visual images — photographs, icons, or signs — rather than either written or spoken words.

Present Information in Appropriate Detail — What level of detail of information does the user need to know? There are many options and the choices can enhance or hinder performance. Users sometimes

Digital displays are often best when precise information is required.

Analog gauges are faster and clearer for giving general indication.

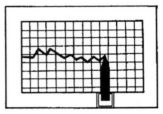

Moving pointers are often best for relative or time-related information.

Warning: This is hard to read. Do not do this.

Warning: This is hard to read. Do not do this.

need only general information; signals should be correspondingly approximate and general. At other times, detailed and precise information is crucial.

Another common example is a training session. Sometimes all that the participants in the training session need is an overview, whereas other participants in other sessions desire in-depth materials. Unfortunately, in training sessions this principle is too often violated: Participants are given unneeded information or an unwarranted level of detail.

The design of signs, instruction manuals, and control panels all can benefit from evaluation. What information does the user need to know? It is the job of the writer/designer to do all the hard work of thinking, organizing, and expressing. The reader/user should have an easy time understanding. Good design leads to ease in use.

Present Clear Images — Another common problem is exhibiting an image poorly so that the user cannot distinguish or interpret the message. Three issues in presenting clear images are being visible, distinguishable, and interpretable. Each of these will be described in more detail.

Make it visible — First, the message must be visible. The size and location should be appropriate at the distance from which it is to be observed and there should be no obstructions. Signs and labels should contrast with their background.

Make it distinguishable — The message should also be distinguishable from other surrounding signals and information. Multiple signals, such as with warning lights or alarms, should not be so similar that they can be confused. For example, a fire alarm should have a pattern or pitch that is distinct from a "process down" alarm. There should be adequate space to separate messages from one another.

Make it interpretable — Finally, the message should be interpretable. An example of one type of issue is avoiding use of characters that look alike — 11, B8, QO — and breaking up long strings — (570) 296-9651. Another type of issue is matching the message with the training of the user.

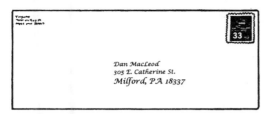

The system for writing addresses provides an example of a useful redundancy — zip code plus the city and state.

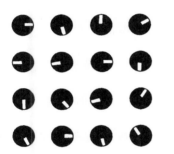

Aligning dials with the zero on top, but indication of normal operation in many different positions, makes it difficult to determine if any one indicator needs attention.

Aligning dials to place the pointers all in the same direction when operations are normal promotes easier detection of abnormal conditions. The eye can quickly perceive the pointer that out of alignment, and thus needing attention.

Use Redundancies — Sometimes, one message is insufficient. Because mistakes are easy to make and humans have many limitations, it is important to provide the same information in more than one way.

A police car that uses both sirens and flashing lights is a common example. Thus, if people miss one cue, they can pick up on another.

The tradition of writing checks with the dollar amounts written out as well as in numeric form helps to overcome bad handwriting and making mistakes. Stop signs at road intersections have three redundancies — color (red), shape (octagonal), and wording (STOP).

The standard system for writing addresses provides an example of a useful redundancy. Technically, with the zip code, the city and state are not needed, since all the information the post office needs is contained in the code. However, it is very easy to make a mistake writing numbers, and the names serve to help the post office correct the error.

Use Patterns — The human eye grasps patterns well. Information presented as a pattern can often be understood much more quickly and accurately than otherwise.

When displaying numerical data, graphs are much easier to read and interpret than columns of numbers. Bar charts are especially good for comparing numbers, and line charts are good for showing trends.

In control panels for complex equipment, grouping appropriate controls can ease use. Similarly, placing controls in patterns or in a context can help indicate to the user what to do. Conversely, using patterns that differ from expectations can be confusing.

Provide Variable Stimuli — Humans detect a novel stimulus more readily than a constant one because our senses fatigue easily with continuous exposure. For example, flashing lights are easier to spot than lights that stay the same. Buzzers that sound only infrequently are noticed more than are recurrent ones. Thus, it is important to avoid excessive use of a single way of presenting information.

Methods of providing warnings serve as examples in other ways. Written warnings are notorious for being ignored, since they quickly become part of the background and simply are not read. The detailed safety instructions on commercial airlines that are provided verbatim for every flight are seemingly not heard by experienced travelers. Only when the information is provided in a unusual way (such as a video featuring an unconventional character) do the frequent fliers pay attention. Some emergency vehicles have been designed to exploit this principle — the sirens switch patterns from time to time to evoke attention.

Applications can affect many arenas. Training sessions that switch from one format to another can be more effective than unremitting exposure to a single method of presentation.

Provide Instantaneous Feedback — An additional principle that helps prevent errors is to provide feedback to the user on the course of action taken. Furthermore, the sooner the feedback is given, the easier it is to determine if an error has been made or not.

Pilots signal that a message has been received with the word *Roger* (a term itself chosen because these low frequency sounds were more easily understood over a static-ridden intercom in a propeller aircraft than, for example, a simple "yes" with its high frequency sounds).

When taking a telephone message, repeating numbers helps clarify that the correct message is being conveyed. The keys on computer keyboards are deliberately designed to click to help indicate that a character has been successfully transmitted.

The total absence of feedback prevents even knowing if a mistake has been made or not, and it greatly increases the likelihood that an error will be repeated.

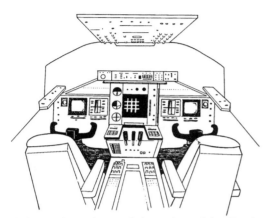

Aviators have learned the value of feedback to acknowledge that commands and communications have been received.

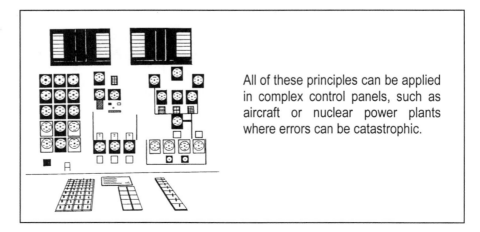

All of these principles can be applied in complex control panels, such as aircraft or nuclear power plants where errors can be catastrophic.

Case Example – Legibility and Ease in Reading

STUDIES HAVE BEEN DONE THROUGH THE YEARS TO DETERMINE THE MOST READABLE TYPEFACES. WRITING IN ALL CAPITAL LETTERS IS NOT ADVISABLE FOR TEXT. WHILE CAPITAL LETTERS CAN BE USED WITH GREAT EFFECTIVENESS IN TITLES OR WHEN THERE IS A NEED TO HIGHLIGHT CERTAIN WORDS, USING ALL CAPITALS IN A BLOCK OF TEXT MAKES IT DIFFICULT TO READ. ONE PARTICULAR PROBLEM IS THAT IT BECOMES MORE DIFFICULT TO TELL WHERE A SENTENCE BEGINS AND WHERE IT ENDS. WEB PAGES ARE CURRENTLY A MEDIUM WHERE THIS MISTAKE IS OFTEN MADE.

Writing in fancy scripts is also not advisable for text. Fancy scripts can look very pleasing to the eye in short phrases or blocks of texts, such as invitations to a wedding. However, when an entire paragraph or page is written in a fancy script, all the flourishes start to create a messy look and are confusing to the eye. Some of the flourishes, in fact, tend to lead the eye to a different line, rather than leading the eye forward along the same line. In short, fancy scripts can be confusing when used excessively.

This is a *serif* typeface and is usually the best one for text. In a serif typeface, such as Times, the letters have tips designed to give a horizontal flow to the lines, thereby enhancing reading. Take a close look at each of the letters used in this paragraph. Note how each has been designed to give a horizontal structure to the words. The letter "h," for example, has a small horizontal tip at the top and two horizontal tips at the bottom, one for each leg. As another example, the letter "o" has thin tops and bottoms and thick sides. These subtle shapes create a flow that propels the eye forward along the same line.

This is a *sans serif* typeface (*sans* being French for *without,* hence "without a serif"). Helvetica and other kinds of sans serif typefaces often have a cleaner look, and are useful for captions and labels, but they are generally not optimal for long blocks of text. With the absences of any tips in the shapes of the letters, the horizontal flow of the text is lost. Compare this paragraph with the preceding one.

Narrow columns are generally easier to read than long lines of text because the eye can more readily shift back to the next line without skipping to the wrong place. Narrow columns also facilitate taking in a complete phrase at a time.

Newspapers and magazines routinely use columns for this reason. In the nineteenth century, many hardcover books also used columns, a practice that modern publishers have unfortunately discarded. Webpages provide the worst examples of long lines of text that are difficult to read (plus often have interfering backgrounds that that make the text indistinguishable).

13

Work Organization

A final area to be highlighted — if only to mention — is "work organization," which is a topic that refers to the underlying design of work and a huge range of issues:

1. **Job-specific, micro-decisions**, such as task allocation, that are made day-to-day by millions of people.

2. **Various administrative practices,** such as compensation systems, that are planned as part of a business strategy.

3. **Overall management systems** that are often taken for granted as part of the philosophical basis of modern industry.

The term *socio-technical* systems has also been used in this context to describe the intertwined area of organizational structures interrelated with technological development. More recently, *macro-ergonomics* has been used to connote this whole context of work, rather than the micro-details in the design of hand tools, computer keyboards or cockpits. Some researchers use the term *psycho-social* to refer to the human organizational side of the workplace, as opposed to the strictly physical and technical side. All in all, these terms mean much the same thing.

Work Organization is a Human Tool

Throughout history, humans have developed both physical tools and successive ways to *organize* to expand human capabilities and overcome human limitations. In this sense, organization itself is a human tool, and thus a good target for ergonomics.

"Work organization" can be thought of in anthropological sense — the various ways humans have arranged to perform tasks to obtain the necessities of life. We might think, for example, of the division of labor among hunter-gatherers, then how people re-organized this division of labor with agriculture, and so forth into the present. Furthermore, we also might think of all the different ways that various cultures around the world have organized their activities at each stage of this development — different societies have made different choices about how work was to be accomplished.

Task allocation — How should tasks be divided and assigned to accomplish goals? Is it better to have many people equally capable of doing many tasks? Or is it better to have a narrow division of labor so that individuals can be extremely highly qualified at specific tasks?

Assemblyline or work cells — Should the technology and equipment of the workplace be designed so that tasks are narrowly defined? Should the physical layout of a workplace promote team activities, such as by having meeting rooms adjacent each production area?

Shift work — Should there be more than one shift in a given workplace? If so, should employees be assigned to just one shift (thus prohibiting some people from enjoying normal evening family and social activities), or should they be rotated between shifts every couple of weeks (thus forcing everyone to disrupt their biological time clocks).

Reward system — How should people be compensated for their activities? What are the actions that are rewarded? Should people be compensated for how much they put into a task (hours and effort), or how much they put out (quality and quantity of product)?

Structure — How many vertical layers should there be in an organization? What degree of horizontal segmentation? What amount of centralization?

Decision-making — What kinds of decisions should be made at what levels of the organization? Should the strategic issues be left to just top managers? Should rank-and-file employees be allowed — or required — to take part in decision-making?

These are the kinds of organizational issues that are encompassed by the term *work organization*. Again, we can think of these organizational issues themselves as tools. As with any other tool, ergonomics seeks to analyze and improve the fit between the system of organization and the people who make up that organization. In parallel with purely physical issues, the goal here is to design systems that match organizational issues with human requirements and thereby increase (a) overall efficiency of the system and (b) personal fulfillment and well-being of people.

Other fields of study have obviously addressed these issues in great detail, but ergonomics provides a certain perspective. In particular, ergonomics more than other fields of study focuses on the point where both the technological side of production and the human side intermesh. For example, management science and organizational psychology address many of these areas, but they do not always capture the technical side. Conversely, engineering addresses much of this subject, but it does not always capture the human side. The combination of the two is what interests ergonomists — the interrelationships of humans and their tools.

To a great extent technology structures what is possible for organizations, but it certainly does not predetermine every aspect of an organizational system. For example, in recent years we have seen that products such as automobiles do not always need to be built on an assemblyline to be the most efficient method of production. Because the assemblyline technology is there does not mean it always must be used, nor is it always the best way; other methods of organizing work are possible. We may take much for granted — the assumptions made because of how it has "always" been done.

Nonetheless, there are still many options of ways to organize modern industrial work. The value of thinking of organization as a tool is that it puts our organizational systems in the same league as physical items such as software and handtools, and thus factors that can be modified and made more friendly to benefit efficiency and well-being.

A truly innovative company or society holds these organizational issues as fair game for change. Ergonomics can provide the same challenges to assumptions about the organization of work to promote innovation as it has for physical tasks.

Technology is changing constantly, and at a more rapid pace than ever before. There is a crucial need to evaluate the options for organizational structures to best take advantage of the new technology. Addressing work organization opens more doors and raises more opportunities. Choices are being made, and ergonomics can help identify those that are both more efficient and most compatible with human fulfillment and well being.

Addressing all these issues is clearly beyond the scope of this book — indeed, any single book. The point is that the ergonomics perspective of fitting tasks to people can provide insights to these organizational assessments.

Part II

Measurements and Guidelines

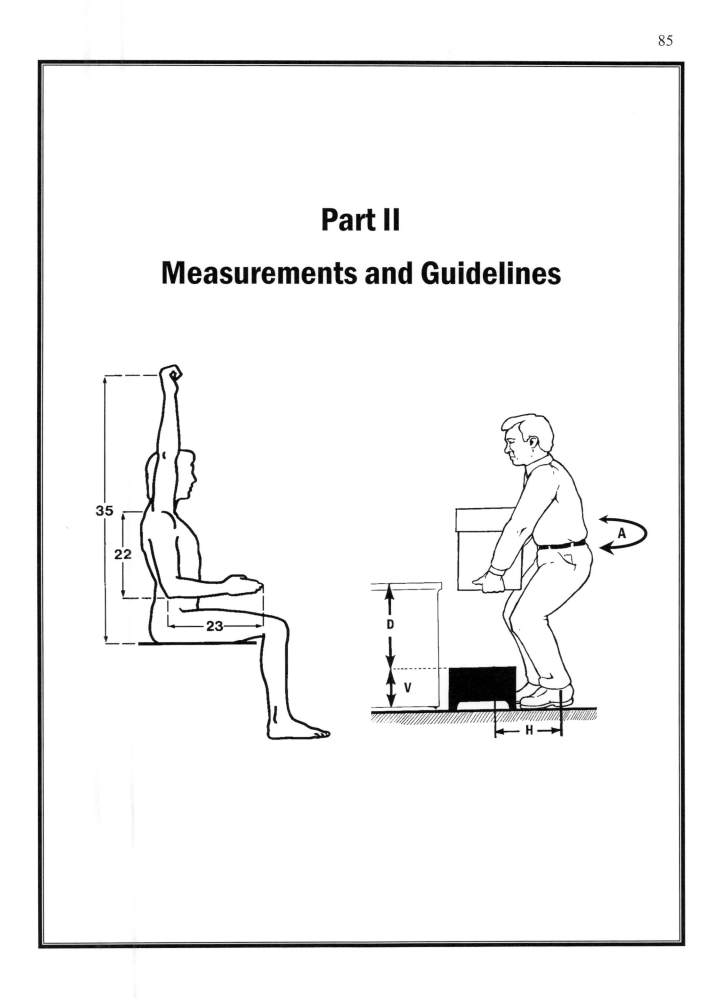

Quantitative Methods

Introduction — Why Measure?

Up to this point, no numbers have been used in this book, for a specific reason — most of the time no measurements are needed to make practical workplace improvements. The process is typically a matter of analyzing tasks in a *qualitative* manner, then following the basic principles, and being creative in their application. The general rule is that it is more important to *think* than to *measure*.

The concept of task analysis is currently the source of considerable confusion in this regard. Specifically, there is a common tendency to equate *analysis* with *measurements*. However, *analysis* simply means taking a whole system and breaking it into its sub-parts (it is the opposite of *synthesis*). In the case of job analysis, it means breaking down a job by tasks and task steps, by body part, and by ergonomics principle. It does not necessarily mean that measurements of any sort must be made.

The confusion can be clarified by applying the phrase, "pick the right tool for the right job." The tools of ergonomics are many and varied — some are basic and some are complex; some are quantitative in nature and others rely more on creative thinking. The technique and approach must fit the needs and goals of the specific workplace. Good task analysis and problem solving can be achieved using simple checklists and worksheets.

However, in the appropriate circumstances, measurements can provide useful, sometimes crucial, insights on tasks. Examples of when such measurements are helpful include:

- <u>Risk assessment</u> — to help evaluate the degree of risk of injury associated with a given task.
- <u>Numeric guidelines</u> — as part of the effort to employ those guidelines that do exist for ergonomics assessment.
- <u>Conducting "before-and-after" studies</u> — to evaluate the effectiveness of job improvements.
- <u>Helping determine best alternatives</u> — to compare and select the best of several options for improvement.
- <u>Matching jobs to employee restrictions</u> — to document existing job requirements so that employee capabilities are not exceeded.
- <u>Evaluating appropriateness for job rotation</u> — to optimize job rotations such that sufficiently different muscle-tendon groups are used to prevent CTDs.
- <u>Conducting epidemiological studies</u> — to compare job conditions with health effects.
- <u>Documenting problems</u> — to help determine if improvements are needed or not, or to assist in setting priorities.

The following materials provide background on some of the more useful measurement tools and available guidelines in the field of physical workplace ergonomics. Note that guidelines for many issues (such as how many repetitive wrist motions are excessive) simply do not yet exist. However, new techniques will undoubtedly be devised in the future that will provide additional value.

Objectives

These materials are designed for the practitioner in the workplace. They describe a variety of quantitative methods and guidelines that can be useful for practical applications in the workplace. There is sufficient information in this document to enable a reader to measure a number of primary variables as well as to apply key guidelines in basic situations. However, there are broader objectives for learning quantitative methods:

- To increase the practitioner's capability for evaluating ergonomics issues *qualitatively*; that is, to sharpen one's ability to see problems and potential improvements.
- To learn about the state of the art of ergonomics methods, their strengths and shortcomings, and to receive a glimpse of more powerful techniques that the future might bring.
- To understand when it may prove useful to obtain support from professional ergonomists and others either to perform measurements in the workplace or conduct more rigorous studies in a laboratory setting.

General Structure of Each Topic

The following general plan will be used for each of the subsequent sections:

- Technical definition and discussion.
- Working tools that are practical for the workplace and can be performed by plant personnel.
- Measurements and evaluations best done in a laboratory situation or by experts.

Each of the sections is identified with the corresponding basic principles of ergonomics. Note, however, that in a number of cases there is not a one-to-one correspondence.

Who is Qualified?

It should be mentioned at the outset that almost anyone with good sense is capable of performing most, if not all, of these basic techniques, especially with practice and proper guidance. One does not need to be a professional ergonomist or engineer to learn these methods.

Indeed, in practice, many production employees and union representatives have been trained to use these tools and have performed accurately. In fact, in some cases, production personnel are better qualified than outside experts because they are more familiar with the tasks and are more able to characterize representative situations. Various professional ergonomists have monitored the performance of hourly personnel, supervisors, and union representatives using these methods and have concluded that with proper support and direction, these individuals are quite capable.

Ergonomics Measurements

There are a number of categories of ergonomics measures. In this book, the focus is on the final set in the following list regarding physical task requirements.

Safety, Health, and Well-Being
- Injury/Illness Rates
- Workers' Compensation
- Discomfort Survey (anonymous "symptoms" survey)
- Active Medical Surveillance (hands-on physical exams)

Human Resource
- Satisfaction
- Absenteeism
- Turnover

Production
- Errors
- Output
- Defects and other quality measures

Physical Task Requirements
- Heights and reaches
- Clearances
- Force
- Posture
- Motions
- Fatigue
- Environmental

14

Anthropometry

Principle 3 — Keep Things in Easy Reach
Principle 4 — Work at Proper Heights
Principle 8 — Provide Clearance

Introduction

Anthropometry is the field of study that deals with the measurement of the human body (*anthropos* means "human" and *metria* means "measurements"). The following pages contain some of the standard human dimensions commonly needed for workplace design, including:

- Stature
- Forward reach
- Upward reach
- Sitting eye height

- Arm span
- Knee clearance
- Thigh clearance
- Shoulder breadth

Anthropometric data is useful in obtaining criteria for the design and evaluation of furniture, tools, and other equipment. For example, the anthropometric tables on the following pages contain many standard dimensions that can be used to determine:
- the range of height adjustability required to accommodate a given population
- the maximum reach acceptable of small individuals
- the minimum clearance acceptable for large individuals.

Pitfalls

Three important factors can affect decisions when choosing and using anthropometric data and interpreting results:

1. **The task.** The nature of the work being done often affects the dimensions. For example, most work should be done at elbow height, which is easily found in the following tables, but there may are usually obtained without regard to clothing. Consequently, one must account as appropriate for specialized be a need to adjust up for precision tasks or down for heavy tasks.
2. **Related tools and products.** Dimensions of any additional tools or products used in the task may also affect your choice of data. For example, the height of a conveyor assembly line should be lowered from "standard" height if the products being assembled are large.
3. **Clothing.** All data clothing, such as hard hats, boots, heels, gloves, and heavy winter clothing.
4. **Workforce differences**. The workforce in a particular facility may differ from the population base from which the data were obtained:
   - Age
   - Gender
   - Ethnic background
   - General, industrial, or military populations

Background — The Normal Curve

Whenever large populations are measured the results tend to fall in a typical pattern — the bell-shaped normal curve. Most people are in the center of the curve. The few people with extreme measurements are at either end — the smallest or weakest on the left and the biggest or strongest on the right.

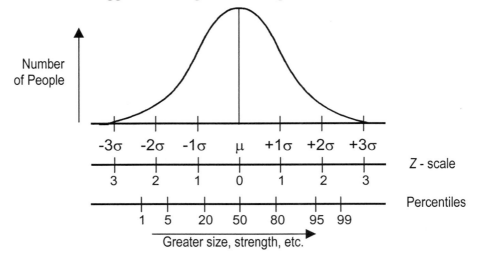

Mean = μ = 50th Percentile

The "average" of this population is known as the "mean," which is often denoted by the Greek letter μ (pronounced "mu"), or sometimes as the "50th Percentile."

Standard Deviation = σ = S.D.

The standard deviation indicates the degree of spread of the particular population away from the mean, often denoted by the Greek letter σ (pronounced "sigma") or by the initials "S.D." Standard deviations are often used in establishing equipment design criteria to accommodate differences in people.

S.D.	Percent of population
± 1	about 68%
± 2	about 95%
± 3	virtually all

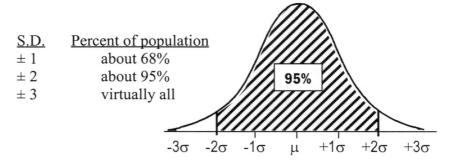

It is convention to establish design criteria at the 5th and 95th percentile (90% of population) or alternatively at ± 2 S.D. (95% of population). The choice of these or any other design criterion depends upon the task at hand and information available. (Note that percentile is not the same as percentage of the population.)

Using Anthropometry Tables

The illustrations on the next page show a number of common measurements that are important in the design of workplaces. To use this section, find the number to the dimension that best reflects the needs at hand, then refer to the corresponding number on the accompanying table.

The data are traditionally segmented by gender. In practice, designers tend to use males for the upper end of a measurement and females for the lower end, thus accommodating more people than if the genders were mixed. Key examples of this concept are:

- For clearance, use large males.
- For reaches, small females.
- For ranges, such as for adjustable workbenches, use both large males and small females.

In the subsequent tables, data are presented both in the percentile format and as mean and standard deviations. Results differ slightly depending upon which is used. Either approach is "correct," but the mean and standard deviation format provides for more inclusion.

For large males, one identifies the mean and adds two standard deviations (or chooses the measurement of 95th percentile male). For small females, one subtracts two standard deviations (or uses the 5th percentile female).

Precision

There are occasions when precision is important, but in most practical applications in the workplace, these measures can be rounded off. For long dimensions such as arm span and overhead reach, using measurements within an inch or so is satisfactory; seeking answers to the fraction of an inch is generally not essential.

The point is to make increasingly informed decisions rather than to guess or to not think about workforce dimensions at all, which has too often been the case in the past. As a final remark, one should almost never simply look up a dimension and use it blindly. These data are to help the thinking process, not replace it.

Additional Reference

The data on these pages are excerpted from Pheasant, 1996. This is an excellent reference book that should be referred to for further details and explanations, including data from additional populations around the globe, infants, the aged, and various additional measures.

A source for additional details and data on more dimensions is Crew System Ergonomics Information Analysis (CSERIAC) which is based on large numbers of U.S. Air Force personnel. There are also a number of additional sources of anthropometric data and tables.

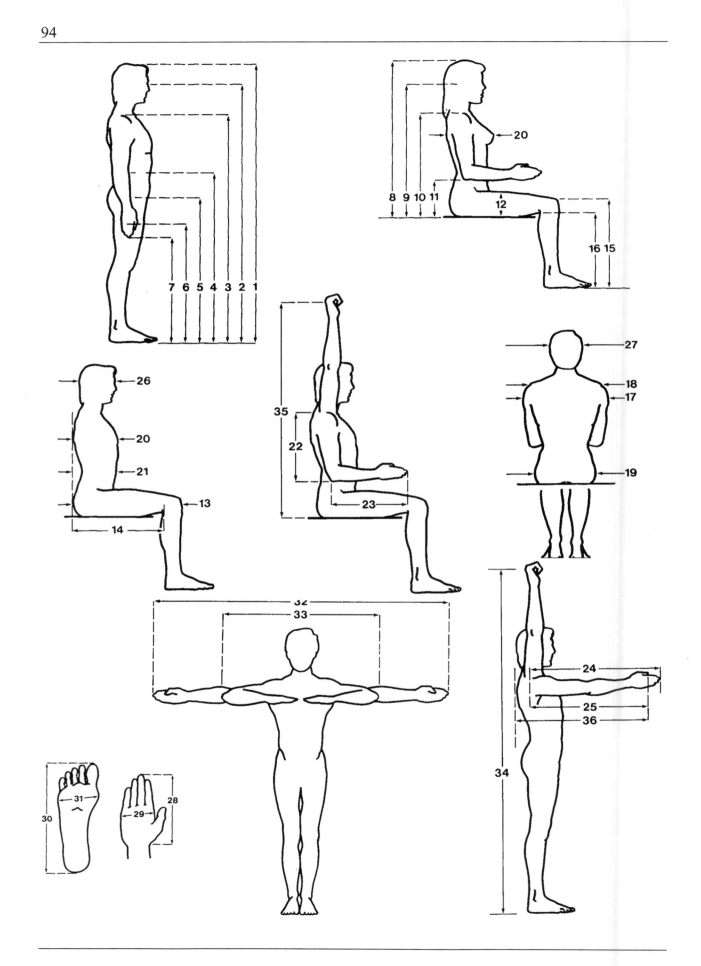

Anthropometry Table
U.S. Adults (Inches)

Measures	Males				Females			
	5th	50th	95th	1 S.D.	5th	50th	95th	1 S.D.
1. Stature	64.6	69.1	73.6	2.8	59.8	64.0	68.1	2.5
2. Eye Height	62.8	67.3	71.9	2.8	55.9	60.0	64.2	2.5
3. Shoulder Height	52.4	56.7	61.0	2.6	48.2	52.2	56.1	2.4
4. Elbow Height	40.2	43.5	46.9	2.1	37.2	40.2	43.1	1.9
5. Hip Height	32.9	36.0	39.2	2.0	29.9	32.9	35.8	1.8
6. Knuckle Height	27.6	30.1	32.7	1.6	26.4	28.7	31.1	1.5
7. Fingertip Height	23.4	26.0	28.5	1.5	22.2	24.8	27.4	1.6
8. Sitting Height	33.7	36.0	38.4	1.4	31.5	33.9	36.2	1.4
9. Sitting Eye Height	29.1	31.5	33.9	1.4	27.2	29.5	31.9	1.4
10. Sitting Shoulder Height	21.5	23.6	25.8	1.3	20.1	22.2	24.4	1.3
11. Sitting Elbow Height	7.7	9.6	11.6	1.2	7.3	9.3	11.2	1.1
12. Thigh Thickness	5.3	6.3	7.3	0.6	4.9	6.1	7.3	0.7
13. Tailbone-Knee Length	21.7	23.6	25.6	1.2	20.7	22.6	24.6	1.2
14. Tailbone-Popliteal Length	17.5	19.7	21.9	1.3	17.3	19.3	21.3	1.2
15. Knee Height	19.5	21.7	23.8	1.3	18.1	19.9	21.7	1.1
16. Popliteal height	15.6	17.5	19.5	1.1	14.2	15.9	17.7	1.1
17. Shoulder Breadth (bideltoid)	16.7	18.5	20.3	1.1	14.2	15.7	17.3	1.0
18. Shoulder Breadth (biacromial)	14.4	15.7	17.1	0.8	13.0	14.2	15.4	0.7
19. Hip Breadth	12.2	14.2	16.1	1.2	12.2	14.8	17.3	1.5
20. Chest (Bust) Depth	8.7	10.0	11.4	0.9	8.3	10.0	11.8	1.1
21. Abdominal Depth	8.7	10.8	13.0	1.3	8.3	10.2	12.2	1.2
22. Shoulder-Elbow Length	13.0	14.4	15.7	0.8	12.0	13.2	14.4	0.7
23. Elbow-Fingertip Length	17.5	18.9	20.3	0.8	15.7	17.1	18.5	0.8
24. Upper Limb Length	28.7	31.1	33.5	1.4	25.8	28.1	30.5	1.4
25. Shoulder-Grip Length	24.2	26.4	28.5	1.3	22.0	24.0	26.0	1.2
26. Head Length	7.1	7.7	8.3	0.3	6.5	7.1	7.7	0.3
27. Head Breadth	5.7	6.1	6.5	0.2	5.3	5.7	6.1	0.2
28. Hand Length	6.9	7.5	8.1	0.4	6.3	6.9	7.5	0.4
29. Hand Breadth	3.1	3.5	3.9	0.2	2.6	3.0	3.3	0.2
30. Foot Length	9.4	10.4	11.4	0.6	8.7	9.4	10.2	0.5
31. Foot Breadth	3.5	3.9	4.3	0.2	3.1	3.5	3.9	0.2
32. Span	65.7	71.3	76.8	3.3	59.3	64.0	68.7	2.9
33. Elbow Span	34.4	37.6	40.7	1.9	31.1	33.9	36.6	1.7
34. Vertical Grip Reach (Standing)	76.8	81.9	87.0	3.1	71.1	75.8	80.5	2.9
35. Vertical Grip Reach (Sitting)	45.5	49.4	53.3	2.4	42.1	45.7	49.2	2.2
36. Forward Grip Reach	28.5	30.9	33.3	1.4	25.8	28.0	30.1	1.3
37. Body Weight (in pounds)	121.0	171.6	224.4	30.8	90.2	143.0	195.8	33.0

Anthropometry Tables
British Adults (Millimeters)

Measures	Males				Females			
	5th	50th	95th	1 S.D.	5th	50th	95th	1 S.D.
1. Stature	1625	1740	1855	70	1505	1610	1710	62
2. Eye Height	1515	1630	1745	69	1405	1505	1610	61
3. Shoulder Height	1315	1425	1535	66	1215	1310	1405	58
4. Elbow Height	1005	1090	1180	52	930	1005	1085	46
5. Hip Height	840	920	1000	50	740	810	885	43
6. Knuckle Height	690	755	825	41	660	720	780	36
7. Fingertip Height	590	655	720	38	560	625	685	38
8. Sitting Height	850	910	965	36	795	850	910	35
9. Sitting Eye Height	735	790	845	35	685	740	795	33
10. Sitting Shoulder Height	540	595	645	32	505	555	610	31
11. Sitting Elbow Height	195	245	295	31	185	235	280	29
12. Thigh Thickness	135	160	185	15	125	155	180	17
13. Tailbone-Knee Length	540	595	645	31	520	570	620	30
14. Tailbone-Popliteal Length	440	495	550	32	435	480	530	30
15. Knee Height	490	545	595	32	455	500	540	27
16. Popliteal height	395	440	490	29	355	400	445	27
17. Shoulder Breadth (bideltoid)	420	465	510	28	355	395	435	24
18. Shoulder Breadth (biacromial)	365	400	430	20	325	355	385	18
19. Hip Breadth	310	360	405	29	310	370	435	38
20. Chest (Bust) Depth	215	250	285	22	210	250	295	27
21. Abdominal Depth	220	270	325	32	205	255	305	30
22. Shoulder-Elbow Length	330	365	395	20	300	330	360	17
23. Elbow-Fingertip Length	440	475	510	21	400	430	460	19
24. Upper Limb Length	720	780	840	36	655	705	760	32
25. Shoulder-Grip Length	610	665	715	32	555	600	650	29
26. Head Length	180	195	205	8	165	180	190	7
27. Head Breadth	145	155	165	6	135	145	150	6
28. Hand Length	175	190	205	10	160	175	190	9
29. Hand Breadth	80	85	95	5	70	75	85	4
30. Foot Length	240	265	285	14	215	235	255	12
31. Foot Breadth	85	95	110	6	80	90	100	6
32. Span	1655	1790	1925	83	1490	1605	1725	71
33. Elbow Span	865	945	1020	47	780	850	920	43
34. Vertical Grip Reach (Standing)	1925	2060	2190	80	1790	1905	2020	71
35. Vertical Grip Reach (Sitting)	1145	1245	1340	60	1060	1150	1235	53
36. Forward Grip Reach	720	780	835	34	650	705	755	31
37. Body Weight (in kilograms)	55	75	94	12	44	63	81	11

Anthropometry Tables
Japanese Adults (Millimeters)

Measures	Males				Females			
	5th	50th	95th	1 S.D.	5th	50th	95th	1 S.D.
1. Stature	1560	1655	1750	58	1450	1530	1610	48
2. Eye Height	1445	1540	1635	57	1350	1425	1500	47
3. Shoulder Height	1250	1340	1430	54	1075	1145	1215	44
4. Elbow Height	965	1035	1105	43	895	955	1015	36
5. Hip Height	765	830	895	41	700	755	810	33
6. Knuckle Height	675	740	805	40	650	705	760	33
7. Fingertip Height	565	630	695	38	540	600	660	35
8. Sitting Height	850	900	950	31	800	845	890	28
9. Sitting Eye Height	735	785	835	31	690	735	780	28
10. Sitting Shoulder Height	545	590	635	28	510	555	600	26
11. Sitting Elbow Height	220	260	300	23	215	250	285	20
12. Thigh Thickness	110	135	160	14	105	130	155	14
13. Tailbone-Knee Length	500	550	600	29	485	530	575	26
14. Tailbone-Popliteal Length	410	470	510	31	405	450	495	26
15. Knee Height	450	490	530	23	420	450	480	18
16. Popliteal height	360	400	440	24	325	360	395	21
17. Shoulder Breadth (bideltoid)	405	440	475	22	365	395	425	18
18. Shoulder Breadth (biacromial)	350	380	410	18	315	340	365	15
19. Hip Breadth	280	305	330	14	270	305	340	20
20. Chest (Bust) Depth	180	205	230	16	175	205	235	18
21. Abdominal Depth	185	220	255	22	170	205	240	20
22. Shoulder-Elbow Length	295	330	365	21	270	300	330	17
23. Elbow-Fingertip Length	405	440	475	20	370	400	430	17
24. Upper Limb Length	665	715	765	29	605	645	685	25
25. Shoulder-Grip Length	565	610	655	26	515	550	585	22
26. Head Length	170	185	200	8	160	170	180	7
27. Head Breadth	145	155	165	7	140	150	160	6
28. Hand Length	165	180	195	10	150	165	180	9
29. Hand Breadth	75	85	95	6	65	75	85	5
30. Foot Length	230	245	260	10	210	225	240	9
31. Foot Breadth	95	105	115	5	90	95	100	4
32. Span	1540	1655	1770	70	1395	1485	1575	56
33. Elbow Span	790	870	950	48	715	780	845	41
34. Vertical Grip Reach (Standing)	1805	1940	2075	83	1680	1795	1910	69
35. Vertical Grip Reach (Sitting)	1105	1185	1265	49	1030	1095	1160	41
36. Forward Grip Reach	630	690	750	37	570	620	670	31
37. Body Weight (in kilograms)	41	60	74	9	40	51	63	7

Anthropometry Tables
Indian Agricultural Workers (Millimeters)

Measures	Males			
	5th	50th	95th	1 S.D.
1. Stature	1540	1620	1700	50
2. Eye Height	1425	1510	1595	52
3. Shoulder Height	1265	1345	1425	49
4. Elbow Height	940	1025	1105	40
5. Hip Height	800	865	930	38
6. Knuckle Height	635	685	730	29
7. Fingertip Height	540	585	630	28
8. Sitting Height	795	840	880	25
9. Sitting Eye Height	695	740	780	26
10. Sitting Shoulder Height	520	555	590	21
11. Sitting Elbow Height	170	05	235	20
12. Thigh Thickness	110	135	160	13
13. Tailbone-Knee Length	520	555	590	21
14. Tailbone-Popliteal Length	435	465	495	18
15. Knee Height	460	510	560	30
16. Popliteal height	380	415	450	21
17. Shoulder Breadth (bideltoid)	375	410	440	19
18. Shoulder Breadth (biacromial)	320	355	395	24
19. Hip Breadth	280	310	335	16
20. Chest Depth	145	170	05	20
21. Abdominal Depth	140	185	235	33
22. Shoulder-Elbow Length	325	355	385	19
23. Elbow-Fingertip Length	425	460	490	20
24. Upper Limb Length	700	755	810	34
25. Shoulder-Grip Length	655	710	760	32
26. Head Length	170	180	190	6
27. Head Breadth	140	145	150	4
28. Hand Length	170	185	195	8
29. Hand Breadth	75	85	90	4
30. Foot Length	235	50	65	10
31. Foot Breadth	90	95	105	4
32. Span	1595	1705	1810	66
33. Elbow Span	825	880	935	33
34. Vertical Grip Reach (Standing)	1875	1995	2110	72
35. Vertical Grip Reach (Sitting)	1120	1190	1265	44
36. Forward Grip Reach	685	725	765	24
37. Body Weight (in kilograms)	40	49	59	6

Anthropometry Tables
Brazilian Industrial Workers (Millimeters)

Measures	Males			
	5th	50th	95th	1 S.D.
1. Stature	1595	1700	1810	66
2. Eye Height	1490	1595	1700	66
3. Shoulder Height	1315	1410	1510	60
4. Elbow Height	965	1045	1120	49
5. Hip Height	800	880	960	47
6. Knuckle Height	655	720	785	40
7. Fingertip Height	565	625	690	37
8. Sitting Height	825	880	940	35
9. Sitting Eye Height	720	775	830	34
10. Sitting Shoulder Height	550	595	645	29
11. Sitting Elbow Height	185	230	275	28
12. Thigh Thickness	120	150	180	16
13. Tailbone-Knee Length	550	595	650	30
14. Tailbone-Popliteal Length	435	480	530	29
15. Knee Height	490	530	575	27
16. Popliteal height	390	425	465	24
17. Shoulder Breadth (bideltoid)	400	445	490	27
18. Shoulder Breadth (biacromial)	355	385	415	19
19. Hip Breadth	305	340	385	25
20. Chest Depth	205	235	275	22
21. Abdominal Depth	220	245	305	33
22. Shoulder-Elbow Length	335	365	405	21
23. Elbow-Fingertip Length	440	475	510	22
24. Upper Limb Length	725	785	850	38
25. Shoulder-Grip Length	615	670	725	34
26. Head Length	175	190	205	8
27. Head Breadth	140	150	160	6
28. Hand Length	170	185	200	9
29. Hand Breadth	75	85	95	5
30. Foot Length	240	260	280	12
31. Foot Breadth	95	100	110	5
32. Span	1625	1755	1885	78
33. Elbow Span	855	925	995	44
34. Vertical Grip Reach (Standing)	1895	2020	2145	75
35. Vertical Grip Reach (Sitting)	1130	1220	1310	56
36. Forward Grip Reach	710	765	820	32
37. Body Weight *(in kilograms)*	52	66	86	11

Interpolations

It is possible to calculate specific measurements based on statistical rules. Values are typically presented at the 5th, 50th and 95th percentiles, along with the measure of S.D. Due to the characteristics of the standard normal distribution, one can determine any percentile, either for determining a range or a specific point. For example, what percentile is a given individual's height?

To determine these values, one can use a statistical tool called the "Z-Table." Z simply represents the number of standard deviations away from the mean.

Therefore, the area between: $\pm 1\ Z = 68.26\%$ of the population
$$\pm 2\ Z = 95.44\%$$
$$\pm 3\ Z = 99.74\%$$

The equation $Z = \dfrac{x_i - \mu}{\sigma}$ is used for Z values that are not whole numbers.

The Z-Table on the following page shows specific percentages for Z values ranging from -3 to +3. Specific points may be read directly from the Z-Table. Determining ranges requires subtracting the upper limit from the lower limit.

The following examples demonstrate the use of the Z-Table:

A. What is the percentile ranking in stature for a 5'7" (67") female?

Step 1: Find μ and σ from the Anthropometry Table, measure #1:
$\mu = 64.8$; $\sigma = 2.8$

Step 2: Calculate the Z statistic: $Z = \dfrac{67 - 64.8}{2.8} = 0.79$

Step 3: Look up percentile of $Z = 0.79$ from Z-Table.

This woman has a percentile of 78.52%.

B. Given a workbench that adjusts from 41 to 47 inches, what percentage of the male population is accommodated? (Assume work is done at standing elbow height.)

Step 1: Find μ and σ for elbow height (measure #4): $\mu = 45.1$; $\sigma = 2.5$

Step 2: Determine Z values for upper and lower adjustments:

$$Z\ upper = \dfrac{47 - 45.1}{2.5} = 0.76 \qquad Z\ lower = \dfrac{41 - 45.1}{2.5} = -1.64$$

Step 3: Look up percentages for upper and lower adjustments from Z-Table:

Upper = 77.64%   Lower = 5.05%

Step 4: Determine range accommodated by subtracting the lower value from the upper value:

77.64 - 5.05 = 72.59%

This workbench accommodates 72.59% of the male population.

The *Z* Table

Z	0	1	2	3	4	5	6	7	8	9
-3	.13	.10	.07	.05	.03	.02	.02	.01	.01	.00
-2.9	.19	.18	.17	.17	.16	.16	.15	.15	.14	.14
-2.8	.26	.25	.24	.23	.23	.22	.21	.21	.20	.19
-2.7	.35	.34	.33	.32	.31	.30	.29	.28	.27	.26
-2.6	.47	.45	.44	.43	.41	.40	.39	.38	.37	.36
-2.5	.62	.60	.59	.57	.55	.54	.52	.51	.49	.48
-2.4	.82	.80	.78	.75	.73	.71	.69	.68	.66	.64
-2.3	1.07	1.04	1.02	.99	.96	.94	.91	.89	.87	.84
-2.2	1.39	1.36	1.32	1.29	1.26	1.22	1.19	1.16	1.13	1.10
-2.1	1.79	1.74	1.70	1.66	1.62	1.58	1.54	1.50	1.46	1.43
-2.0	2.28	2.22	2.17	2.12	2.07	2.02	1.97	1.92	1.88	1.83
-1.9	2.87	2.81	2.74	2.68	2.62	2.56	2.50	2.44	2.38	2.33
-1.8	3.59	3.52	3.44	3.36	3.29	3.2	3.14	3.07	3.00	2.94
-1.7	4.46	4.36	4.27	4.18	4.09	4.01	3.92	3.84	3.75	3.67
-1.6	5.48	5.37	5.26	5.16	5.05	4.95	4.85	4.75	4.65	4.55
-1.5	6.68	6.55	6.43	6.30	6.18	6.66	5.94	5.82	5.70	5.59
-1.4	8.08	7.93	7.78	7.64	7.49	7.35	7.22	7.08	6.94	6.81
-1.3	9.68	9.51	9.34	9.18	9.01	8.85	8.69	8.53	8.38	8.23
-1.2	11.51	11.31	11.12	10.93	10.75	10.56	10.38	10.20	10.03	9.85
-1.1	13.57	13.35	13.14	12.92	12.71	12.51	12.30	12.10	11.90	11.70
-1.0	15.87	15.62	15.39	15.15	14.92	14.69	14.46	14.23	14.01	13.79
-0.9	18.41	18.14	17.88	17.62	17.36	17.11	16.85	16.60	16.35	16.11
-0.8	21.29	20.90	20.61	20.33	20.05	19.77	19.49	19.22	18.94	18.67
-0.7	24.20	23.89	23.58	23.27	22.97	22.66	22.36	22.06	21.77	21.48
-0.6	27.43	27.09	26.76	26.42	26.11	25.78	25.46	25.14	24.83	24.51
-0.5	30.85	30.50	30.15	29.81	29.46	29.12	28.77	28.43	28.10	27.76
-0.4	34.46	34.09	33.72	33.36	33.00	32.64	32.28	31.92	31.56	31.21
-0.3	38.21	37.83	37.45	37.07	36.69	36.32	35.94	35.57	35.20	34.83
-0.2	42.07	41.68	41.29	40.90	40.52	40.13	39.74	39.36	38.97	38.59
-0.1	46.02	45.62	45.22	44.83	44.43	44.04	43.64	43.25	42.86	42.47
-0.0	50.00	59.60	49.20	48.80	48.40	48.01	47.61	47.21	46.81	46.41
0.0	50.00	50.40	50.80	51.20	51.60	51.99	52.39	52.79	53.19	53.59
0.1	53.98	54.38	54.78	55.17	55.57	55.96	56.36	56.75	57.14	57.35
0.2	57.93	58.32	58.71	59.10	59.48	59.87	60.26	60.64	61.03	61.41
0.3	61.79	62.17	62.55	62.93	63.31	63.68	64.06	64.43	64.80	65.17
0.4	65.54	65.91	66.28	66.64	67.00	67.36	67.72	68.08	68.44	68.79
0.5	69.15	69.50	69.85	70.19	70.54	70.88	71.23	71.57	71.90	72.24
0.6	72.57	72.91	73.24	73.57	73.89	74.22	74.54	74.86	75.17	75.49
0.7	75.80	76.11	76.42	76.73	77.03	77.34	77.64	77.94	78.23	78.52
0.8	78.81	79.10	79.39	79.67	79.95	80.23	80.51	80.78	81.06	81.33
0.9	81.59	81.86	82.12	82.38	82.64	82.89	83.15	83.40	83.65	83.89
1.0	84.13	84.38	84.61	84.85	85.08	85.31	85.54	85.77	85.99	86.21
1.1	86.43	86.65	86.86	87.08	87.29	87.49	87.70	87.90	88.10	88.30
1.2	88.49	88.69	88.88	89.07	89.25	89.44	89.62	89.80	89.97	90.15
1.3	90.32	90.49	90.66	90.82	90.99	91.15	91.31	91.47	91.62	91.77
1.4	91.92	92.07	92.22	92.36	92.51	92.65	92.78	92.92	93.06	93.19
1.5	93.32	93.45	93.57	93.70	93.82	93.94	94.06	94.18	94.30	94.41
1.6	94.52	94.63	94.74	94.84	94.95	95.05	95.15	95.25	95.35	95.45
1.7	95.54	95.64	95.73	95.82	95.91	95.99	96.08	96.16	96.25	96.33
1.8	96.41	96.48	96.56	96.64	96.71	96.78	96.86	96.93	97.00	97.06
1.9	97.13	97.19	97.26	97.32	97.38	97.44	97.50	97.56	97.62	97.67
2.0	97.72	97.78	97.83	97.88	97.93	97.98	98.03	98.08	98.12	98.17
2.1	98.21	98.26	98.30	98.34	98.38	98.42	98.46	98.50	98.54	98.57
2.2	98.61	98.64	98.68	98.71	98.74	98.78	98.81	98.84	98.87	98.90
2.3	98.93	98.96	98.98	99.01	99.04	99.06	99.09	99.11	99.13	99.16
2.4	99.18	99.20	99.22	99.25	99.27	99.29	99.31	99.32	99.34	99.36
2.5	99.38	99.40	99.41	99.43	99.45	99.46	99.48	99.49	99.51	99.52
2.6	99.53	99.55	99.56	99.57	99.59	99.60	99.61	99.62	99.63	99.64
2.7	99.65	99.66	99.67	99.68	99.69	99.70	99.71	99.72	99.73	99.74
2.8	99.74	99.75	99.76	99.77	99.77	99.78	99.79	99.79	99.80	99.81
2.9	99.81	99.82	99.82	99.83	99.84	99.84	99.85	99.85	99.86	99.86
3	99.87	99.90	99.93	99.95	99.97	99.98	99.98	99.99	99.99	100.0

Entries opposite 3 are for 3.0, 3.1, 3.2, etc.

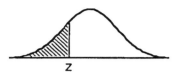

Entries shown in the table represent the cumulative percentage of the population included at or below that Z-value.

For example, when Z=0 (i.e., the mean), 50% of the population lies in the shaded region.

To determine a range (say between ±1 S.D.; i.e., between Z= -1 and Z= +1), first find the upper limit (Z=1 → 84.13), then the lower limit (Z= -1 → 15.87) and subtract (84.13 – 15.87 = 68.26). See the preceding page.

15

Exertion and Biomechanics

Principle 2 — Reduce Excessive Forces

Introduction — Force Versus Exertion

Exertion is the tension produced by muscles and transmitted through tendons to produce force. *Force* is the externally observable result of a specific movement or exertion.

This distinction is important for two reasons. The first is the issue of mechanical advantage; due to the inherent structure of the musculoskeletal system, exertion levels are often many times larger than externally observable forces. This factor contributes directly to musculoskeletal injury. Second, people of different physical sizes may produce the same (external) force to accomplish a task, but may experience different (internal) muscle tension.

For most purposes, exertion is the key factor of concern; however, it is difficult to observe and measure. Thus, it is more common to concentrate on force and assume a correlation with exertion. In this sense, force is a proxy for exertion, recognizing that effects on individuals may well differ.

Several common approaches to measuring force can be accomplished by in-house personnel at the workplace. Additional, more sophisticated techniques can be performed in a laboratory.

Pushing and Pulling

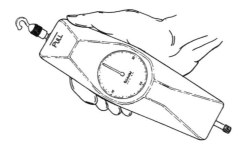

Force Gauge

Some tasks involve lifting or push/pull forces on the arms and shoulders, such as pulling a product off a conveyor line or pushing a cart. These forces are easily measured with a *force gauge*. Force gauges come with a variety of attachments suitable for different tasks and equipment. Furthermore, it is also quite easy for a maintenance department to fabricate special use attachments. Creativity allows maximum usage of these types of tools.

These gauges can be used to measure both peak and maintenance forces, although it is more typical to focus on peak forces. Both are needed for the existing push/pull guidelines discussed later. Output is typically presented on a display in pounds. Since a force gauge is relatively inexpensive and easy to use, it is a good tool for an introductory ergonomics program.

Advanced Techniques

A variety of other devices and techniques have been investigated for characterizing exertion, including electromyography (EMG), load cells, force platforms, and other strain gauges. More information on these methods is described later.

Grasping and Pinching

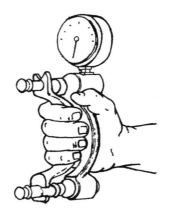

Pinch Gauge and Hand Dynamometer

A simple way to estimate hand exertion is to use either a *pinch gauge* (for the thumb and fingers) or a *hand dynamometer* (for full hand).

<u>Steps</u>

1. Have one or two employees squeeze the device with the same amount of exertion they normally use to do the task. For example, they would attempt to reproduce the effort used to pull a staple gun on the hand dynamometer.

2. Record the result, have them do the task again to remind them of how much effort the task requires, then try the gauge once again.

3. Repeat several times and average the results. The number of samples required depends upon the use of the data and the variance observed.

4. It may be valuable to sample several operators from varying shifts.

This technique approximates the force involved and is dependent on the employee's ability to mimic the effort. Nonetheless, useful measurements can be obtained in this fashion. Both devices are relatively inexpensive and good for an introductory ergonomics program. Medical departments and clinics often have hand dynamometers for use in strength testing and employee screening, which can be borrowed for this purpose.

Advanced Techniques

Two additional methods for measuring grasping forces are (a) EMGs and (b) force sensing resistors. The latter are strain gauges that respond with varying electrical resistance when compressed. As an example of their use, force sensing resistors can be placed on either the surface of a tool or the palm of the hand to determine the force distribution across the hand.

Lifting

Characterizing forces placed on the lower back during lifting tasks is an area that has been well studied. Techniques include biomechanical models (see following pages) and the guidelines for lifting that are described later. Loads to be lifted can easily be measured with a normal scale.

Electromyography (EMG)

An *electromyogram* (EMG) is record of electrical activity in a muscle group. The EMG may be processed numerically and thus is an objective way to characterize muscle exertion. Electrical activity within the muscles is detected by electrodes that are taped on the skin over the muscles in question. EMG testing is primarily done in a laboratory setting, although portable units can be brought to the workplace. This section provides a basic overview to show potential applications for this technique, since an EMG would not normally be used by any workplace personnel.

The following illustrations show the use of an EMG in evaluating a task. In this case, electrodes were placed on the shoulders of an employee. In the awkward situation at left, the product orientation is too high. Consequently, the employee's shoulders are hunched and her elbows extended. In the situation with the neutral arm posture at right, the work has been lowered and properly oriented. Her shoulders are now relaxed and her elbows are at her side.

EMG results are shown below each illustration. Notice in the awkward situation the increased muscle activity of the shoulder, reflected in the graph of the EMG and its numerical evaluation.

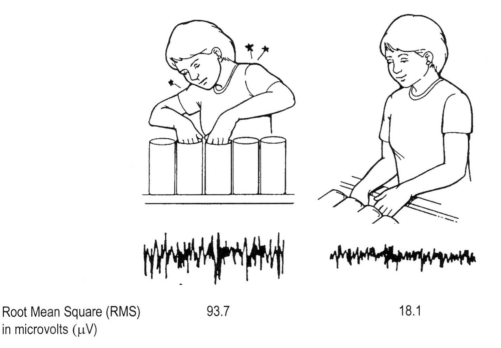

Root Mean Square (RMS)   93.7   18.1
in microvolts (µV)

Pocket EMGs

Small commercially available units can be used in some circumstances by lay people for a rough indication of exertion. These units typically provide auditory output — a beeping sound — or are connected with a voltmeter for a ballpark visual display. A common off the job use for this hand-held device is to provide biofeedback while learning to relax. In the workplace, it can be used to compare different work techniques for some types of tasks, plus help employees learn to relax their grip on tools.

Grip Strength Anthropometric Data

Anthropometric data provide another tool useful for design purposes, such as establishing a criterion for the force required to close a squeeze-operated tool. For occasional squeezing, a useful approach is to use the grip strength of a 5th percentile female. For static gripping, such as for a deadman's grip on a lawnmower, a common practice is to take 15% of that amount (based on Rohmert's Curve, as described later).

Pounds

Age	Male		Female	
	Mean	S.D.	Mean	S.D.
20-24	121.0	20.6	70.4	14.5
25-29	120.8	23.0	74.5	13.9
30-34	121.8	22.4	78.7	19.2
35-39	119.7	24.0	74.1	10.8
40-44	116.8	20.7	70.4	13.5
45-49	109.9	23.0	62.2	15.1
50-54	113.6	18.1	65.8	11.6
55-59	101.1	26.7	57.3	12.5
60-64	89.7	20.4	55.1	10.1
65-69	91.1	20.6	49.6	9.7
70-74	75.3	21.5	49.6	11.7
75+	65.7	21.0	42.6	11.0
Average	**104.3**	**28.3**	**62.8**	**17.0**

Kilograms

Age	Male		Female	
	Mean	S.D.	Mean	S.D.
20-24	266.2	45.3	154.9	31.9
25-29	265.8	50.6	163.9	30.6
30-34	268.0	49.3	173.1	42.2
35-39	263.3	52.8	163.0	23.8
40-44	257.0	45.5	154.9	29.7
45-49	241.8	50.6	136.8	33.2
50-54	249.9	39.8	144.8	25.5
55-59	222.4	58.7	126.1	27.5
60-64	197.3	44.9	121.2	22.2
65-69	200.4	45.3	109.1	21.3
70-74	165.7	47.3	109.1	25.7
75+	144.5	46.2	93.7	24.2
Average	**229.5**	**62.3**	**138.2**	**37.4**

(Mathiowetz, et al., 1985).

Biomechanical Models

Biomechanical models are mathematical predictive equations that can be used to estimate forces. These models are based on principles of physics and mechanics that govern the workings of a "biological machine" just as they would a mechanical one.

Back-of-the-Envelop Calculations

Understanding the basics of biomechanics can enable workplace personnel to make many reasonable, simple calculations. For example, to estimate the force on the wrist from holding a hammer, simply multiply the lever arm by the load to calculate the moment:

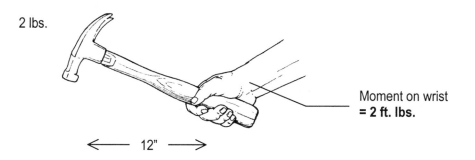

2 lbs.

Moment on wrist
= **2 ft. lbs.**

12"

To take a slightly more complex example, it is similarly possible to estimate the amount of biceps exertion that is needed to lift a weight.

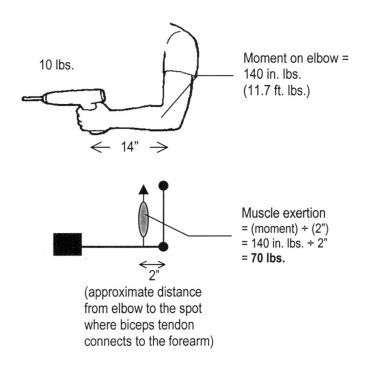

10 lbs.

Moment on elbow =
140 in. lbs.
(11.7 ft. lbs.)

14"

Muscle exertion
= (moment) ÷ (2")
= 140 in. lbs. ÷ 2"
= **70 lbs.**

2"
(approximate distance
from elbow to the spot
where biceps tendon
connects to the forearm)

Back Compression Force

Increasing the level of sophistication, the following is a basic biomechanics model (Bloswick, 1999) for estimating the compression forces on the lower back. This particular approach takes into account three different types of forces:

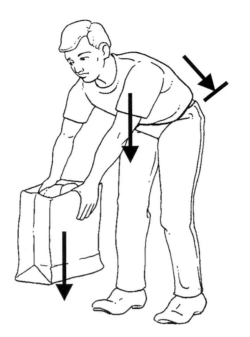

Fc = Compression Force in lower back = $A + B + C$,

where:

$A =$ back muscle force reacting to upper body weight $\quad = 3 \, (BW)\cos(theta)$

$B =$ back muscle force reacting to load $\quad = .5(L*HB)$

$C =$ direct compressive component of upper body weight $\quad = .8[(BW)/2 +L]$

In order to make an estimate of the load on the back, the following data are required:

BW = Body Weight (in pounds)

L = Load (in pounds)

HB = Hand to Back distance (in inches)

Cos(theta) = Torso angle with horizon:

torso posture:	use:
vertical	0
bent 1/4 of the way	.38
bent 1/2 of the way	.71
bent 3/4 of the way	.92
bent horizontal	1.00

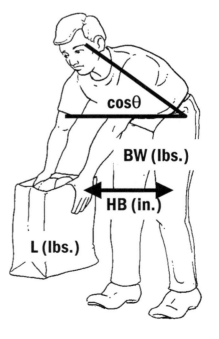

Back Compression Force Worksheet

BW = Body Weight = _____ lbs.

L = Load in Hands = _____ lbs.

HB = Horizontal Distance from Hands to Low Back = _____ in.

Cos(theta) = Torso Angle with Horizon = _____

torso posture:	use:
vertical	0
bent 1/4 of the way	.38
bent 1/2 of the way	.71
bent 3/4 of the way	.92
bent horizontal	1.00

Fc = *A + B + C*

A = 3 (BW)cos(theta) = 3(_____) * (_____) = _____

B = .5(L*HB) = .5(_____) * (_____) = _____

C = .8[(BW)/2 +L] = .8[(_____)/2 + _____] = _____

Total Compression Force Estimate = *A+B+C* = _____ lbs.

Guideline

As a rough indication of the impact of various loads, compression forces above 750 lbs. will put *some* portion of the workforce at risk and compression forces above 1500 lbs. will put *most* members of the workforce at risk.

Uses and Limitations

The prime reason for reproducing this model here is to increase the understanding of how such methods work, thus laying the background for understanding the use of more complicated and accurate computerized models. However, this method can be used to calculate rough estimates of the effect of lifting loads and is simple enough to work by hand using a calculator.

The information on these two pages is excerpted from training materials developed by Don Bloswick, Ph.D., Department of Mechanical Engineering, University of Utah. For a complete explanation of this tool, it applications and limitations, and derivation of the formula, contact him. A similar basic model for estimated loads on the shoulder is also available.

Computer Models

The preceding simple methods can be helpful on occasion. For more accurate needs, the equations become more complex and computers are necessary to perform the calculations.

These more-advanced models take into account an increased number of variables. Furthermore, they include a number of relationships between different linked body segments that permit estimation of, for example, the load on the knees generated by a weight held in the hand.

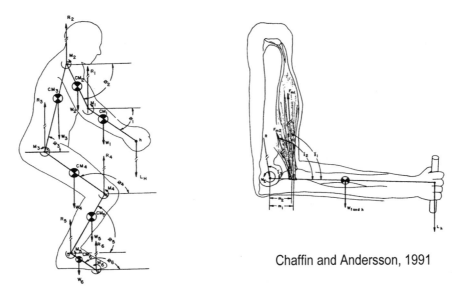

Chaffin and Andersson, 1991

A considerable amount of research and development has gone into the creation of easy-to-use software to apply these models. A common and useful example is a program that can be used to (a) estimate forces occurring in the lower back and (b) estimate strength requirements based on different whole body postures and various simulated tasks. In general, these programs work along the following lines.

Input

The working screen of the software presents a stick figure or other image of a person. The analyst specifies the postural geometry of the stick figure to emulate actual work postures by entering joint angles into the program.

Data on the task are also entered:
- Direction of the force, to account for lifting, pushing, or pulling.
- The amount of the load (obtained with a force gauge or by weighing the object).
- One versus two hand use.
- Work population specifications, such as anthropometric percentiles or gender.

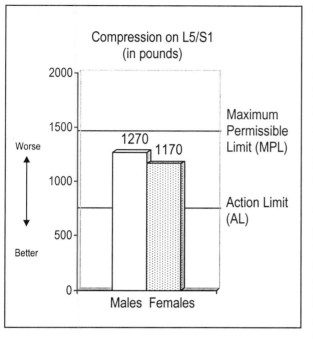

The output is generally provided in bar graphs as well as in tables of calculations. Typically, one graph shows the predicted compression force on the L5/S1 disc of lower back for both males and females for the simulated task. The graph shows limits for comparison, such as the Action Limit (AL) and the Maximum Permissible Limit (MPL). The design goal is to *reduce* the height of these graphs as much as possible.

The second set of graphs typically shows the percentage of the male and female population that have the strength to perform that simulated task for each identified joint or body link. In this case, the design goal is to *increase* the height of these graphs as much as possible, ideally so that 90 or 95% of the population is able to perform the task.

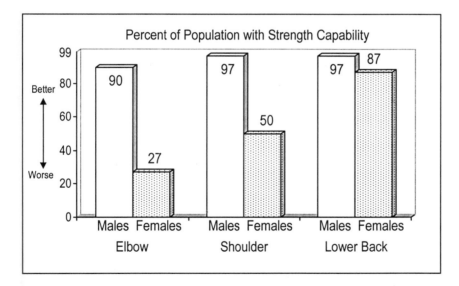

Interpretation

In this example, a load of 80 pounds was specified to be lifted with two hands in a particular posture using the general population. The results show that the compression force on the lower back for males is roughly 1300 pounds, which exceeds both the Action Level (AL) and verges on the Maximum Permissible Level (MPL). The equivalent compression force for females is slightly lower.

Additionally, the resulting graphs show that 90% of the male population and 27% of the female population have the elbow strength to perform the task as specified. Similar data are provided for the shoulder and the lower back.

Comments

Note that for a given situation, males experience greater compression in the lower back than females, due to greater male stature. Additionally, notice that most men have the *strength* to make this lift and thus predispose themselves to injury.

The preceding example shows a two-dimensional (2D) version of the software. Versions with three dimensional (3D) capabilities are also available.

As a final comment, remember that results based on the use of biomechanics are estimates, not actual measurements. The magnitude of compression and exertion in the various segments of the body is based on geometry and mechanical relationships, like a structural engineer would view the span of a bridge.

Design Goals

The design goals when interpreting these graphs are the following:

- Compression Force — Change loads and postures to reduce compression forces as much as possible, preferably below the Action Limit (i.e., the bars should decrease as much as possible).

- Strength Capability — Change the loads and postures to permit as large a percentage of the workforce as possible to do the task safely (i.e., the bars on the graph should increase).

3D and the Future

These programs are relatively simple to run, once one gains an initial understanding of how the software works. They can easily be used by engineers and other personnel in the workplace and should be counted as a tool that is available for workplace use, and not just for the ergonomics research laboratory.

These biomechanical models currently have limitations based on assumptions that may not be realistically reflected under actual working conditions. In particular, they focus on static postures, although most work involves movement, with changes in load momentum and consequently varying forces on the body. However, the software can provide value with the right application and interpretation. Furthermore, software of this type is rapidly increasing in sophistication with increased computer capabilities and additional available data. Some versions of these programs integrate anthropometric data with the strength data, plus are compatible with various CAD programs. Future versions hold considerable prospect for useful applied tools in the workplace.

16

Posture

Principle 1 — Work in Good Postures

Characterizing Posture

Introduction

Posture is characterized by measuring the angular relationship between various body links and a fixed frame of reference. The most intuitive frame of reference is another major body segment, such as the forearm for the posture of the wrist or the torso for the posture of the neck. However, some systems use other types of references, such as the horizon. Measures include both the magnitude and the duration of the specific postures.

Two common approaches evaluating postures are (a) visual checklists which are simple to use, but provide less information, and (b) software programs which can characterize a posture much more accurately, yet still remaining relatively simple to use.

Visual Checklist

The following are examples of the checklist approach for the arms and the rest. The technique is simply to observe the joint in question on the person at the task, refer to the pictures below in determining which posture category applies, and mark the category that indicates the most typical posture. This approach obviously works best for relatively static tasks or when the concern is for a specific job step or action. However, it is possible to use a checklist to record two postures, one performed infrequently and one performed frequently. When this occurs, it is possible to record both by labeling the frequent posture **F** and the infrequent posture **I**.

Dominant-Arm Posture	__At Side	__<45° From Side	__45-90° From Side	__90-135° From Side
Non Dom-Arm Posture	__At Side	__<45° From Side	__45-90° From Side	__90-135° From Side

at side

<45° from side

45° - 90° from side

90° to 135° from side

Analyzing wrist posture is often the hardest part of posture analysis. Hand motions are easy to miss because they often occur quickly. The pictures below depict common types of wrist postures.

Dominant Wrist Posture	__Flex>20°	__Extend>20°	__Radial Dev.	__Ulnar Dev.	__Torquing	__N/A
Non Dom Wrist Posture	__Flex>20°	__Extend>20°	__Radial Dev.	__Ulnar Dev.	__Torquing	__N/A

Flexion   Neutral   Extension

Radial Deviation   Neutral   Ulnar Deviation

The following procedure can be helpful in determining hand and wrist postures:

<u>Steps</u>

1. Concentrate on one hand at a time. Start with the dominant hand and look only for *flexed wrist* postures. Watch several cycles until it is determined if the hand is ever flexed more than 20°.

2. Then, using the same procedure, concentrate on *wrist extension* greater than 20°.

3. Next, using the same procedure, concentrate on *radial deviation* in the wrist.

4. Finally, using the same procedure, concentrate on *ulnar deviation*.

5. Watch for circular motions of the wrist. Examples of torquing are wringing clothes or using a screwdriver. The key is to watch the forearm near the wrist. The hand or arm will often flip over when torquing occurs.

6. After evaluating these postures in the dominant hand, move on to the non-dominant hand and repeat steps one through five.

Computer Posture Analysis

Software is available to help conduct more sophisticated and detailed analyses. The software is quite simple to use. With these data, jobs can easily be compared, such as to study alternative work methods. Steps are:

1. Videotape a representative portion of the task.

2. While watching the completed video, record the posture as the joint moves. For example, one system involves holding down one of several keys that have been programmed to indicate the posture in question. In other words, on the computer keyboard, the letter *F* for example would represent Neutral, *G* represent Flexion >45%, *H* represent Flexion >90°, and *J* represent Other. Thus, during the course of the task, one or another of these keys is held down to characterize the postures that are observed.

3. Repeat as necessary to study all necessary joints.

Sample Printout		
Joint	Posture	Percent of Time
Shoulder	Neutral	45
	Flexion >45°	39
	Flexion >90°	10
	Other	6
	Total	100

Electric Goniometer

A goniometer is a device used to measure joint angles. Simple goniometers are similar to protractors. Electric goniometers have been developed and can be used in the workplace. A hinge-like device is placed on the joint in question, which detects the changes in angle and transmits the data to a computer. Results can easily be displayed both graphically and numerically. This method is the most accurate of those described here, but is also the most expensive and difficult to use. It would normally be expected that experts, rather than workplace personnel, would use this technique.

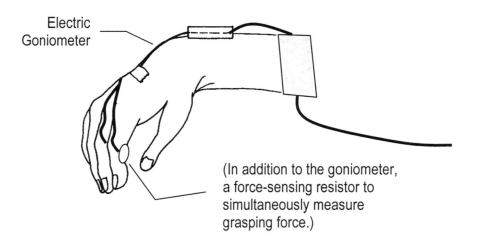

Electric
Goniometer

(In addition to the goniometer,
a force-sensing resistor to
simultaneously measure
grasping force.)

Nomenclature System

To aid evaluation of postures, terminology has been developed to describe the various postures. Using this system, virtually every movement and position of the body can be described.

Normally, these terms are not needed for day-to-day activities that focus on engineering improvement of jobs. However, the terms are sometimes used as part of medical management as well as in studies and various reports. Thus, the workplace practitioner should have some familiarity with at least a few of the basic terms.

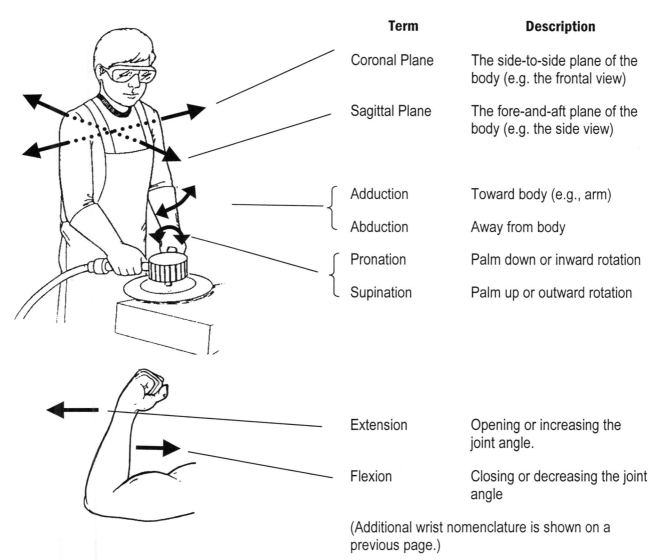

Term	Description
Coronal Plane	The side-to-side plane of the body (e.g. the frontal view)
Sagittal Plane	The fore-and-aft plane of the body (e.g. the side view)
Adduction	Toward body (e.g., arm)
Abduction	Away from body
Pronation	Palm down or inward rotation
Supination	Palm up or outward rotation
Extension	Opening or increasing the joint angle.
Flexion	Closing or decreasing the joint angle

(Additional wrist nomenclature is shown on a previous page.)

17

Motions

Principle 5 — Reduce Excessive Repetition

Motions and Repetitions

A *motion* is a movement or exertion made by a major joint or body link. Usually, the concern is for the number of loads on, or slides of, a tendon and is thus the primary criterion for quantifying the motions of a task. The rate of acceleration of the motion may also be an important factor, altho at present beyond the scope of practical workplace activities.

Motions are quantified simply by observing and counting, either directly or by viewing a video-tape. A good rule of thumb is that a motion occurs every time the body part in question changes directions. Thus, the extremely common flexion/extension cycle involves two motions.

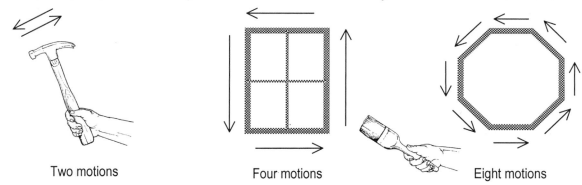

Two motions   Four motions   Eight motions

Counting motions can often be a very simple undertaking, since it can be done by the unaided eye and counting to oneself. On the other hand, counting can sometime be quite difficult, since it can be troublesome to determine exactly what constitutes a motion in many tasks. Further-more, as of this printing, there is little standardization in the field and different analysts tend to use different definitions and systems. Widespread concern for the effects of motions on em-ployee well being is, all in all, a recent event and, prior to now, there has been no particular need for standardization.

Current differences in terminology include the following:

- The term *repetition* is the word commonly used in day-to-day language in the context of risk factors for CTDs. Although "repetition" has a useful connotation when thinking generally about CTDs, it is imprecise and creates a mental stumbling block when it comes to quantifi-cation.

- *Repetition* is commonly used in ways that mean both "motions and "a series of interrelated motions." For example, the Lifting Guide described later involves counting the number of lifts to characterize the repetitiveness of a task. However, one *lift* under that framework would usually consist of two motions using the definition here.

- A repeated sequence of motions is commonly referred to as a *cycle*. A lift as described in the preceding paragraph could be described as a cycle. However, the matter is yet further com-plicated since a particular series of lifts could also be described as a cycle. Some analysts re-fer to both cycles and sub-cycles in this context.

- Jobs that involve high numbers of motions may have few cycles, thereby falsely appearing to involve relatively few motions. For example, consider a maintenance job that requires thou-sands of hand and arm motions per day, but has few repetitive cycles. The untrained observer therefore might seriously underestimate the number of motions involved, since it is not a classical repetitive job like an auto assembler or meatpacker.

In these materials, the focus is placed on motions, as determined by the change in direction of a moving body part. Cycles are irrelevant except as an aid in counting motions per day, as described shortly. The term *repetition* should be avoided in the context of quantification.

To be sure, it may still be important to report results in different ways because of these various interpretations. For example, if the purpose of counting is to help determine tasks that are suitable for restricted duty work, medical personnel reading a report may assume that a "motion" is a "cycle," and the physical requirements of the task might then appear unreasonably high. Consequently, many ergonomists attempt to resolve this whole dilemma by capturing as much detail as possible, and then synthesizing the results in such a manner that the data provide the most meaning for the person reading the report.

As a final comment, the goals for making the measurements greatly affect the precision and definitions of the motion. For example, making a before-and-after estimate of a task improvement requires considerably less precision than a scientific study to associate work factors with injury of joints and tissue.

Counting Techniques

Steps

1. Identify the joint or body link in question when observing a task — the wrist, elbow, or back, left versus right, and so on. Focus only on that specific joint and ignore everything else, guarding against becoming distracted by other movements and the task itself.

2. Videotape the task whenever possible. Videos can (a) help the analyst to focus, (b) enable use of slow motion to detect rapid motions, such as is common with the wrist, and (c) permit multiple reviews of the same sequence or different sequences to help increase accuracy. Background investigation may be necessary to help insure that the segment of work being studied is representative of normal operations.

3. Identify the steps of the job. It is often useful to document the motions for each step in the process of obtaining a count for the entire task.

4. Select the appropriate counting strategy, either based on production output or sampling a representative work cycle (see later).

5. Count the number of times the body link in question changes direction.

   a. Include exertions, even if no movement is detected. For example, attempting to lift a car with a single effort would be considered a motion, even tho the car was not moved.

   b. Ignore wiggles and minor movements of the joint that are less than ±15°, particularly if they are unrelated to the task. (This range of ±15° is a reasonable, but arbitrary, number that might vary depending upon the joint in question. Some analysts use ±10° and others ± 20°.)

There can be exceptions to this general rule. Options at the analyst's choice are:

    i. Include minor movements if there is an exertion on the joint, especially if major movements are unloaded.

    ii. It may be appropriate for some tasks to count 10 minor motions (an arbitrary number) as one normal motion.

    iii. Include minor movements if there is some other reason for including them.

  c. Count complete circular motions as four motions.

  d. Torque motions should be counted separately

6. Document these specific definitions and the assumptions made for the task or workplace at hand. Many choices are arbitrary, but once made, should become a standard that is followed. Furthermore, the documentation permits others to interpret the results.

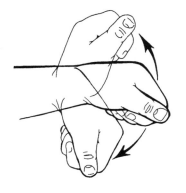

Strategies

Product Output Method

The product output method is the preferred technique and may be used when the rate of work is controlled by machine output, moving conveyors, or known quantity of daily output. This method is especially important when the goal is to determine how many motions a person experiences during a full work shift, since that reflects the total daily "dose." To calculate the motions using the product output method:

<u>Steps</u>

1. Count the number of motions per piece. (It is often helpful to count several representative cycles, then average the results.)

2. Determine the product output of the machine (or down the conveyor, etc.) in terms of pieces per hour or pieces per day. In general, use *maximum* production rates if the goal of the analysis is to estimate job requirements for medical purposes such as restricted duty. Use *average* production rates for other purposes, such as before-and-after studies.

3. Multiply the number of motions per piece by the product output. This provides the overall motions per hour or per day.

4. If more than one employee is involved in the task, divide the overall count by the number of persons performing the task.

Example: Calculate motions/hour when packaging widgets into a box for shipment using the product output method.

1. 16 wrist motions are required in the right hand to package widgets (including grasping box, packing six widgets, and sealing the box).
2. The maximum output is 320 packaged boxes per hour.
3. 16 x 320 = 5,120 motions per hour.
4. 3 people perform this job.
5. 5,120/3 = 1,700 motions per hour per person.

Sampling Method (Snapshot)

If the workflow varies over time and is dependent on the employee (self-paced), the sampling method offers the next best technique. This approach is not as accurate as the product output method since a single snapshot of a task may not take into account variations occurring throughout the day. If the goal of the measurement is to reflect the daily dose, selecting random times for sampling can help eliminate bias. Alternatively, if the goal is to compare methods or conduct a before-and-after study, then selecting a specific task or time period is suitable.

Steps

For a representative one-minute period:

1. Count the number of motions that occur.

2. Repeat several times, then average results.

When using this method, it is important not to extrapolate directly from this number. Simply multiplying this rate by the number of minutes and hours in a day can exaggerate the estimated peak (or minimum) sampled rates or ignore the effect of lunch and rest breaks. A good practice is to keep careful notes on what factors were accounted for and to do one's best to understand how other measurements were taken to insure that comparisons are appropriate.

Time Frame

A common goal is to determine how many motions a person is exposed to during a full work shift, since that reflects the total daily "dose." However, it is also normal to document results using motions per minute or motions per hour. The value in doing so is usually to reflect peak (or sometimes minimum) exposures.

18

Miscellaneous

Principle 6 — Minimize Fatigue
Principle 7 — Minimize Contact Stress
Principle 9 — Move, Exercise, and Stretch
Principle 10 — Maintain a Comfortable Environment

Fatigue

Metabolic Load

Heart Rate Monitor — A good measure of overall exertion is heart rate, which can easily be measured with simple heart- rate monitors. This standard technique is well based in science and is particularly effective since the monitors are small, lightweight and do not interfere with the employees' normal work activities. The heart rate is measured with a small patch that is worn across the chest. This patch transmits heart rate data to a small receiver that can be worn like a wristwatch or pinned to the clothing. Data from the receiver is uploaded daily into a computer and later analyzed. There are accepted guidelines for heart rate levels and exertion and this technique can, for example, help define the relationship between workload and staffing requirements for physically demanding jobs.

Static Load

Time — Static load is measured in terms of time. The classical perspective is that of *the number of minutes* it takes before the onset of pain. More recently, with the concern for highly repetitive jobs, the approach taken has been to study *the percentage of time* in which a single muscle group is used. (Note that awkward postures and exertion can worsen the effect of static load, but the primary index of concern is time.)

EMG — Electromyography can also be used to characterize fatigue. The development of local muscle fatigue has been correlated to changes in the frequency composition of EMG activity. Observing these changes requires sophisticated instrumentation and analytic techniques.

Contact Stress

The unit of measurement for contact stress is pounds per square inch. However, measurements of this sort in applied situations are not common.

Discomfort surveys (see later) have occasionally been used specifically to measure contact stress, such as in the evaluation of different types of floor matting.

Movement and Exercise

Exercise physiologists and kinesiologists have developed elaborate systems for measurement of various types of motions and exercise. However, these techniques are not commonly used for applied workplace issues.

Environment

Vibration

A common measurement technique is attaching sensors called *accelerometers* either to the body (i.e., the palm of the hand) or to the vibrating tool or piece of equipment. A particularly useful device is a hand/arm vibration meter. Exposure limits have been developed for vibration, although they are oriented for Reynaud's Phenomenon and not Carpal Tunnel Syndrome.

Temperature

A simple measure of temperature can be taken with a standard thermometer. A more accurate approach is to use a Wet Bulb Globe Thermometer (WBGT), which takes the following factors into account:

- Radiant heat — effect of external bodies such as the sun or a furnace
- Convective heat — gain or loss from exposure to moving air
- Evaporation — heat loss due to evaporation of moisture

Cognition

During the past few decades a set of measurement techniques has been developed to quantify cognitive issues. The methods have been refined in both the aerospace industry and in the field of experimental psychology. They typically measure items such as:

- Reaction time — how long it takes to react to a certain stimulus
- Performance time — how long it takes to complete a task
- Quality of output — how many errors were made while doing the task
- Quantity of output — how many items in total were completed
- User reactions — how the subject felt, measured with opinion ratings, preference scales and the like.

Differences in design of various types of products can be evaluated through experimentation or review of actual work situations. Two examples of techniques used to analyze these issues are the usability lab and the mental workload assessment.

Usability Lab

A "Usability Lab" is a unique technique that has been developed to study human error. A product is placed in a room and human subjects are instructed to use it. Observers watch to see what mistakes are made and to record comments. In a full lab, a one-way mirror is typically used to separate the observers from the subjects, plus the actions are all videotaped. The resulting observations can be studied to determine either typical mistakes or items that the subjects found self-evident and easy to use. Thus, designers are given feedback and guidance on layouts and configurations.

Mental Workload Assessment

A common technique for measuring mental workload is called *Secondary Task Measures*. Subjects are given a primary task (the real task to be evaluated, such as an air traffic controller's job), plus a secondary task (an unrelated arbitrary task, such as having to hit a button every third time a light blinks). The primary task is not measured directly; rather, it is evaluated by the response to a secondary task; that is, the number of correct responses. When a person is overloaded on the primary task, the score on the secondary task is easily affected.

Work Organization

Effectiveness and Satisfaction

The impact of various types of work organization systems can be measured in a wide variety of ways, including measurements of productivity, quality, job satisfaction, and overall profitability. Work organization is a huge topic and it has been addressed in rigorous ways by researchers in the fields of engineering, business administration, and organizational psychology.

Stress

One specific issue in work organization that has implications for ergonomics is stress. The physiological stress response is complex reaction of the body to a threat — muscles tense, heart rate increases, and the body prepares to defend itself. The reaction can occur as a result of threatening occurrences in daily life, such as time pressures for completing a task or a confrontation with another person. Stress can be measured in two ways: (a) self-administered questionnaires and (b) physiological measurements such as adrenaline excretion.

19
Lifting Guideline

The NIOSH Lifting Guide

Background

The National Institute for Occupational Safety and Health (NIOSH) is part of the U.S. Department of Health and Human Services and is charged with research and education. In contrast, the Occupational Safety and Health Administration (OSHA) is part of the U.S. Department of Labor and is charged with enforcement.

In 1981, NIOSH published *The Work Practices Guide for Manual Lifting*. The criteria used to establish this guideline included a review of biomechanics, physiology, and the most pertinent studies of back injuries. In 1991, NIOSH drafted a second edition of the guideline, on which the following information is based. Although there are many limitations to this guideline, it provides an approach more useful than simplistic weight limits of the past, which overlooked many of the crucial variables in lifting.

Risk factors associated with manual lifting

The following are factors that contribute to overexertion injuries associated with manual lifting:

Load-Related Factors	**Environmental Factors**	**Personal Factors**
Weight of Load*	Temperature	Age
Horizontal location of load*	Humidity	Gender
Vertical location of load*	Lighting	Anthropometry
Vertical distance of lift*	Noise	Physical fitness and training
Frequency and duration of lifting*	Vibration	Lumbar mobility
Twisting while lifting*	Foot traction	Medical history
Grip (slippery, handles, etc.)*	Posture constraints	Years of employment
Stability of load	Obstacles	Personal health habits
Distance carried	Work organization issues	
Size of load		

Sufficient information is known about the factors indicated by an asterisk (*) to estimate their impact on safe lifting, and they are included in the NIOSH Lifting Guideline. These factors can easily be measured at the worksite and directly evaluated by the guideline. The remaining risk factors are not directly incorporated into the guideline. They are, however, recognized as issues, but assumed in the Lifting Guide to be optimal. The guideline does not directly take into account the individual performing the task. Instead, several personal risk factors data were normalized into the coefficients of the lifting guideline as fixed values representing worker capacity.

The following materials provide an overview of the Lifting Guide that permit use in assessing elementary lifts; however, it is best to consult the full guide and background information (NIOSH, 1991) for active use. Note that understanding the concept of the Lifting Guide — how multiple variables can be incorporated to determine a limit for a specific task — is as important as actual use of the equation.

For basic applications, use the following steps:

Step One — Review the task in question and measure the following variables, using a tape measure and a watch. (All distances in this explanation are in inches; the metric formula follows.)

1. Determine if the lift requires controlling the load at the destination, like loading parts into a fixture of a machine. If it does, then, measure both origin and destination values for both horizontal and vertical distances and for angle of twist (see subsequent worksheet).

2. Measure the **Horizontal distance** between the hands and the midpoint between the ankles at the start and end of the lift. This distance is known as **H**. Restrictions: If H is less than 10, then set H equal to 10. If H is greater than 25, then the lift should be redesigned so that H is 25 or less.

3. Measure the **Vertical distance** of the hands from the floor at the start and end of the lift. This distance is known as **V**. Restrictions: V should be a value between 0 and 70 — the floor and the upper limit of vertical reach. If V is greater than 70, then the lift should be changed so that V is 70 or less.

4. Measure the **vertical travel Distance** from the start of the lift to the destination. This distance is known as **D**. Restrictions: If D is less than 10, then set D equal to 10. Since the upper limit of the lift is 70, D should also be less than 70.

5. Estimate the **Angle of twist** of the body in degrees measured at the beginning and end of the lift. This angle is known as **A**. Restrictions: A should be a value between 0° and 135°. If A is greater than 135°, then the lift should be redesigned so that A is 135° or less.

6. Count the average **Frequency of lifting** in lifts per minute. Additionally, determine the **duration** of lifting (<1 hrs/shift, < 2 hrs/shift, or < 8 hrs/shift). Refer to the Frequency Table to determine the frequency multiplier, known as **F**. Restrictions: If F is less frequent than once per 5 minutes (.2 lifts/min.), then set F equal to .2 lifts/min.

7. Evaluate the **hand-to-container Coupling** used during the lift. Refer to the Coupling Table to determine the coupling classification, known as **C**.

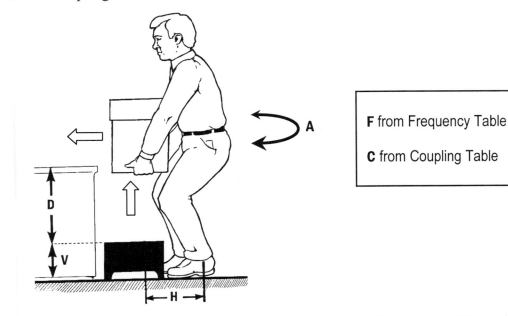

F from Frequency Table

C from Coupling Table

The Equation:

$$RWL = 51 \left[\frac{10}{H} \right] \left[1 - (.0075 \times |V\text{-}30|) \right] \left[.82 + \frac{1.8}{D} \right] \left[F \right] \left[1 - .0032\,A \right] \left[C \right]$$

Step Two — Insert the measurements into the preceding equation to determine **the Recommended Weight Limit (RWL)**. If control is required at the destination, then make two separate calculations: one for the origin and one for the destination, then use the smallest RWL as the acceptable limit.

Concept:

The "51" in the equation is 51 pounds. Note that each factor that is applied reduces the acceptable limit.

Example:

H = 12"

V = 10"

D = 20"

A = 15°

F = 5 lifts per minute for less than 2 hours = .60 (from Frequency Table)

C = Fair Coupling = .95 (from Coupling Table)

then the formula is evaluated as follows:

$$RWL = 51 \left[\frac{10}{12} \right] \left[1 - (.0075 \times |10\text{-}30|) \right] \left[.82 + \frac{1.8}{20} \right] \left[.60 \right] \left[1 - .0032\,(15) \right] \left[.95 \right]$$

$$= 51 \left[.83 \right] \left[.85 \right] \left[.91 \right] \left[.60 \right] \left[.95 \right] \left[.95 \right]$$

$$= \textbf{17.73 lbs}.$$

Note that V = 50" yields the same results.

It is also common to express results as an index compared with the actual weight lifted. This is a convenient way to show results, since anything greater than one (1) calls for action, and it permits comparisons of different lifts. For example, if the weight in the preceding case were 25 pounds:

$$\text{Lifting Index (LI)} = \frac{\text{actual load}}{\text{RWL}} = \frac{25 \text{ lbs.}}{17.73 \text{ lbs.}} = 1.4$$

Lifting Guide Worksheet

Department   _____

Task   _____

Evaluator   _____

Object Weight   _____pounds

      (average   _____pounds)

      (maximum   _____pounds)

H$_o$ _____inches   (**H**orizontal distance to grip at **origin**)

Vo   _____inches   (**V**ertical distance to grip at **origin**)

(If control at destination is required:)

     Hd   _____inches   (**H**orizontal distance to grip at **destination**)

     Vd   _____inches   (**V**ertical distance to grip at **destination**)

D   _____inches   (vertical travel **D**istance)

Ao   _____degrees   (**A**ngle at **origin** - bird's eye, shoulders/feet)

Ad   _____degrees   (**A**ngle at **destination** - bird's eye, shldrs/feet)

F   _____lifts/minute

Duration   _____hours/day   (choose the largest value that is true: <1, <2, <8)

Coupling   _____   (select: good, fair, or poor)

Coupling Definitions	Optimal Container	Optimal Handles
Good	Yes	Yes
Fair	Yes	No
Poor	No	No

Step Three — Determine if any of the following conditions apply.

Lift
____ awkward or unsmooth lifting motion
____ one-handed lift

Load
____ Unstable
____ Excessively large size (e.g., larger than 30 inches)

Environmental
____ Constraints on posture, obstacles in path, poor clearance, and so on.
____ Unusual heat or cold
____ Slippery floors

Personal
____ Employee in less than normal physical shape, condition, strength, or flexibility.
____ Employee with existing back problem or other precondition.

If any of these factors are present then the assumptions of the guide have been violated. Results may overestimate the weight that is acceptable to lift and should be interpreted accordingly.

Step Four — Workstation changes — If the actual weight lifted exceeds the calculated **RWL**, then reconsider the variables measured (H, V, D, etc.).

____ Can the horizontal distance between the hands and the body centerline be reduced?

____ Can the vertical location between the hands and the floor be made closer to waist height?

____ Can the vertical distance between the origin and destination of the lift be reduced?

____ Can the layout be changed to reduce the amount of twist required?

____ Can the way the load is held be improved (better handles, etc.)?

____ Can the repetition of lifts per minute be reduced or the duration of lifting be reduced?

____ Can the load be reduced, either through smaller load weight or use of a mechanical lifting aid?

Complex Tasks
If loads are lifted from varying positions (such as from a full pallet), then each tier and row must be evaluated and results averaged. If multiple lifting tasks are involved, then each must be evaluated and results combined (see the full NIOSH documentation).

Interpretation of RWL

The RWL is defined as "the weight of the load that nearly all healthy workers could perform over a substantial period of time (e.g. up to 8 hours) without an increased risk of developing lifting-related lower back" for a specific set of task conditions.

In the author's judgment, the RWL is a quite stringent and easy to exceed in normal workplace practice. The guideline is probably accurate from a physiological point of view, but it is probably not feasible to meet in many workplaces. In other words, if the goal were to attempt to prevent *all* back injuries, one would want to adhere to the guideline; however, in the real world it would be difficult — perhaps impossible — to do so.

It is helpful to note that the 1981 equation and the first publication of the 1991 equation did not refer to the RWL. Rather the terminology was that of an Action Limit (AL) and a Maximum Permissible Limit (MPL), with the AL identical to the RWL and the MPL set as three times the AL. In many ways, the AL/MPL nomenclature provides a more intuitive understanding of implications, since it provides a range:

- The AL provides an estimate of the level at which *some* workers may experience increased risk of back injuries. Clearly, many workers can lift loads above this level without increased risk.

- The MPL provides an estimate of upper limits, beyond which the risk of injury is seriously elevated for nearly all workers.

For complete background, refer to the NIOSH documentation (NIOSH 1991).

Metric Version

$$RWL = 23 \left[\frac{25}{H}\right] \left[1 - (.003 \times |V\text{-}75|)\right] \left[.82 + \frac{4.5}{D}\right] \left[F\right] \left[1 - .0032\,A\right] \left[C\right]$$

RWL in kilograms
H,V,D in centimeters
F,A,C as previous

Frequency Multiplier Table (FM)

Frequency	Work Duration (Continuous)					
	<8 HRS		<2 HRS		<1 HOUR	
Lifts/min	$V<30$	$V>30$	$V<30$	$V>30$	$V<30$	$V>30$
0.2	.85	.85	.95	.95	1.00	1.00
0.5	.81	.81	.92	.92	.97	.97
1	.75	.75	.88	.88	.94	.94
2	.65	.65	.84	.84	.91	.91
3	.55	.55	.79	.79	.88	.88
4	.45	.45	.72	.72	.84	.84
5	.35	.35	.60	.60	.80	.80
6	.27	.27	.50	.50	.75	.75
7	.22	.22	.42	.42	.70	.70
8	.18	.18	.35	.35	.60	.60
9	.00	.15	.30	.30	.52	.52
10	.00	.13	.26	.26	.45	.45
11	.00	.00	.00	.23	.41	.41
12	.00	.00	.00	.21	.37	.37
13	.00	.00	.00	.00	.00	.34
14	.00	.00	.00	.00	.00	.31
15	.00	.00	.00	.00	.00	.28
>15	.00	.00	.00	.00	.00	.00

Values of V are in inches.

Hand - to - Container Coupling Classification

Couplings	$V<30$ inch	$V>30$ inch
	Coupling Modifiers	
Good	1.00	1.00
Fair	0.95	1.00
Poor	0.90	0.90

GOOD

1. For containers of optimal design, such as some boxes and crates, a "Good" hand-to-object coupling would be defined as handles or handholds of optimal design (see notes 1 to 3).

2. For loose parts or irregular objects, which are not usually containerized, such as castings, stock, or supply materials, a "Good" hand-to-object coupling would be defined as a comfortable grip in which the hand can be easily wrapped around the object (see note 6).

FAIR

1. For containers of optimal design, a "Fair" hand-to-object coupling would be defined as handles or hand-hold cut-outs of less than optimal design (see notes 1 to 4).

2. For containers of optimal design with no handles or hand-holds or for loose parts or irregular objects, a "Fair" hand-to-object coupling is defined as a grip in which the hand can be flexed about 90° (see note 4).

POOR

1. Containers of less-than-optimal design with no handles or handholds or loose parts or irregular objects that are bulky or hard to handle (see note 5).

NOTES:

1. An optimal handle design has 0.75 – 1.5-inch diameter, >4.5-inch length, 2-inch clearance, cylindrical shape, and a smooth, non-slip surface.

2. An optimal handhold has >3-inch height, 4.5-inch length, round shape, 2-inch clearance, smooth non-slip surface, and >0.43-inch container thickness.

3. An optimal container design has <16-inch frontal length, <12-inch height, and a smooth non-slip surface.

4. A worker should be capable of clamping the fingers at nearly 90° under the container, such as required when lifting a cardboard box from the floor.

5. A less than optimal container has a frontal length >16-inch, height <12-inch, rough or slippery surface, sharp edges, asymmetric center of mass, unstable contents, or require gloves.

Computer Spreadsheet

It is best to learn the use of the Lifting Guide by doing it manually. However, for day-to-day use, computer software is normally used. Software is especially important when evaluating a job that involves several different lifting tasks, which is laborious and time consuming to do by hand.

Several commercial versions of the Lifting Guide are available. Furthermore, formatting a spreadsheet or programming a calculator is not difficult.

Multi-Task Job Analysis Worksheet

Date

Department Molding   **Job Description**

Job Title Lift from Extractor   Remove casting from high table on platform

Analyst Dan MacLeod   Carry to separate table on floor

STEP 1: Measure and Record Task Variables

Task #	Object Wt. (lbs)		Hand Location (in.)				Vert. Dist.	Twist(deg.)		Freq.	Dur.	Coupl.	STRWL	More
	L(avg)	L(max)	Ho	Vo	Hd	Vd	D	Ao	Ad	F		C	(lbs.)	Limit.
1	40	45	10	47	0	0	4.5	0	0	0.2	8	fair	37.82	Origin
2														
3														
4														
5														

STEP 2: Compute multipliers and FIRWL, STRWL, FILI, and STLI for Each Task

Task #	HF	VF	DF	AF	CF	FF	x FIRWL =	STRWL	FILI	STLI	New #	Hazard Assessment FILI>1	STLI>1	FILI>STLI
1	1.00	0.87	1.00	1.00	1.00	0.85	44.50	37.82	1.0	1.06	1	**FIX**	**FIX**	**OK**
2														
3														
4														
5														

Note: Abbreviations in this spreadsheet are:

HF	Horizontal Factor	⎫
VF	Vertical Factor	
DF	(vertical) Distance Factor	⎬ (Knowing these helps to identify possible improvements; that is, which factors could likely have the biggest effect.)
CF	Coupling Factor	
FF	Frequency Factor	⎭

STRWL   Single Task Recommended Weight Limit

FIRWL   Frequency Independent Recommended Weight Limit

FILI   Frequency Independent Lifting Index

STLI   Single Task Lifting Index

20

Pushing, Pulling, and Carrying Guides

Pushing, Pulling, and
Carrying Guides

Guidelines have also been developed for other tasks that involve the lower back used in much the same manner as the Lifting Guideline. Software versions of these guidelines are also available.

Steps

1. Review the task and measure the following variables:
   - The **Distance** that the load will be moved.
   - The **Frequency** the task is performed
   - The **Force to Start** the load in motion (using a force gauge).
   - The **Force to Keep** the object in motion.

2. Consult the table:
   - Select the most appropriate column for the **Distance** and the **Frequency** of the task.
   - Select the row for the type of action
   - The value indicated where the column and row intersect is the maximum acceptable force or weight for the task.
   - Interpolate as necessary.

Notes

Ninety percent of the female population should be able to perform at this level without excessive strain.

The data assume the load is handled at about waist height.

The table is excerpted and adapted from Snook and Ciriello, 1990. For additional parameters such as different heights and segments of the population, see the original source. For similar guidelines under different circumstances, such as acceptable forces while kneeling or lying down, see Mital et. al, 1993.

Other Tasks Performed — The tables assume that other tasks performed during the job cycle are minor. If they are not, then load limits should be reduced.

Temperature — The tables assume normal temperature. A hot work environment can significantly reduce a worker's capacity to work. In studies, hot work environments (WBGT 80.6°F) reduced the workload for lifting by 20%, for pushing by 16%, and for carrying by 11%.

Type of Study

These data were developed based on the subjects' subjective responses to the loads that they believed they could handle safely. This type of study is known as "psychometric."

Maximum Acceptable Force

(Pounds)

7 ft.

Distance / One action per	6 Sec.	12 Sec.	1 Min.	2 Min.	5 Min.	30	1 Hr
Push Initial Force	31	33	37	40	44	46	48
Push Sustained Force	13	15	20	20	22	24	29
Pull Initial Force	31	35	40	42	46	48	51
Pull Sustained Force	13	20	22	22	24	26	31
Carry	26	31	35	35	35	35	48

25 ft.

Distance / One action per	10 Sec.	16 Sec.	1 Min.	2 Min.	5 Min.	30	1 Hr
Push Initial Force	31	33	35	37	42	42	46
Push Sustained Force	13	15	18	18	20	20	24
Pull Initial Force	31	33	35	37	42	42	46
Pull Sustained Force	15	18	20	20	22	22	29
Carry	26	26	31	31	31	34	42

50 ft.

Distance / One action per	18 Sec.	24 Sec.	1 Min.	2 Min.	5 Min.	30	1 Hr
Push Initial Force	24	29	31	31	35	35	37
Push Sustained Force	11	13	13	15	15	18	22
Pull Initial Force	22	26	31	31	35	37	40
Pull Sustained Force	11	13	15	15	18	20	24
Carry							

(Kilograms)

2 m

Distance / One action per	6 Sec.	12 Sec.	1 Min.	2 Min.	5 Min.	30	1 Hr
Push Initial Force	14	15	17	18	20	21	22
Push Sustained Force	6	7	9	9	10	11	13
Pull Initial Force	14	16	18	19	21	22	23
Pull Sustained Force	6	9	10	10	11	12	14
Carry	12	14	16	16	16	16	22

8 m

Distance / One action per	10 Sec.	16 Sec.	1 Min.	2 Min.	5 Min.	30	1 Hr
Push Initial Force	14	15	16	17	19	19	21
Push Sustained Force	6	7	8	8	9	9	11
Pull Initial Force	14	15	16	17	19	19	21
Pull Sustained Force	7	8	9	9	10	10	13
Carry	12	12	14	14	14	15	19

15 m

Distance / One action per	18 Sec.	24 Sec.	1 Min.	2 Min.	5 Min.	30	1 Hr
Push Initial Force	11	13	14	14	16	16	17
Push Sustained Force	5	6	6	7	7	8	10
Pull Initial Force	10	12	14	14	16	17	18
Pull Sustained Force	5	6	7	7	8	9	11
Carry							

Snook and Ciriello, 1990

21

Employee Discomfort Surveys

Employee discomfort can be measured through use of self-administered questionnaires. Employees typically are asked to rate any discomfort they might experience on the job, for example, on a scale of one to five. The questionnaires are usually simple one or two page forms that are administered anonymously. Results can be summarized, statistically analyzed, graphed, and evaluated. The purpose of this type of survey is to gain an overview of a particular department, operation, or area. An example of a survey form is shown below:

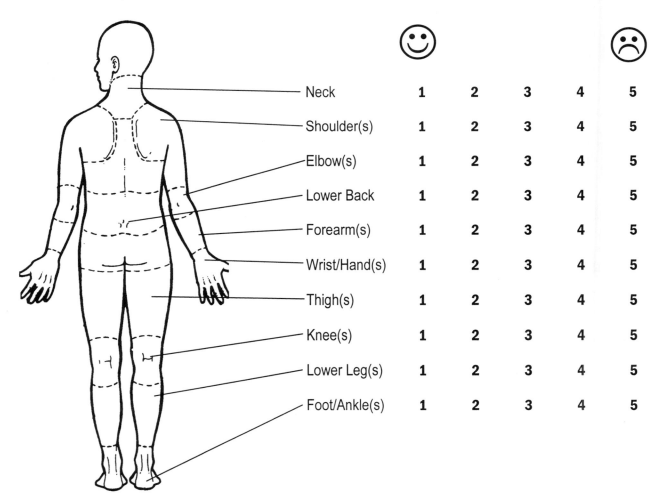

	☺				☹
Neck	1	2	3	4	5
Shoulder(s)	1	2	3	4	5
Elbow(s)	1	2	3	4	5
Lower Back	1	2	3	4	5
Forearm(s)	1	2	3	4	5
Wrist/Hand(s)	1	2	3	4	5
Thigh(s)	1	2	3	4	5
Knee(s)	1	2	3	4	5
Lower Leg(s)	1	2	3	4	5
Foot/Ankle(s)	1	2	3	4	5

Discomfort Survey Results

Two examples of final results are show below. The first relates to a department of 22 employees where the surveys were administered about six months apart to evaluate the effectiveness of ergonomic changes. The second was administered to a smaller group of employees every 45 minutes, again to evaluate the effectiveness of improvements.

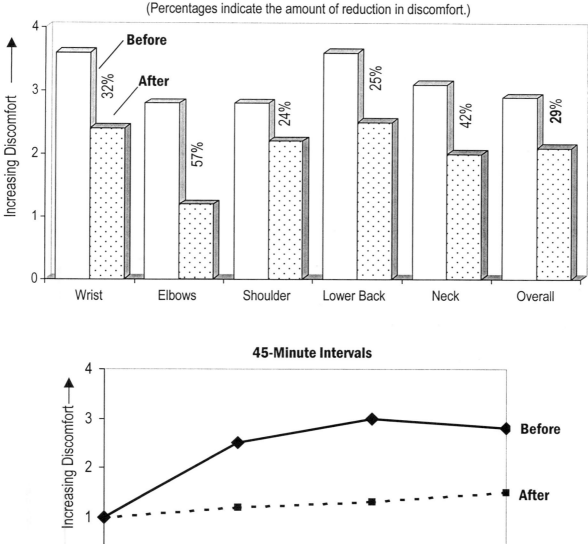

Six-Month Interval
(Percentages indicate the amount of reduction in discomfort.)

45-Minute Intervals

Uses of Discomfort Surveys

Results from discomfort surveys can provide useful insights on potential problems, or at the very least, can confirm what analysts have observed. The same kind of information can also be obtained from one-on-one discussions with employees in the area, but the anonymous survey has its special value, particularly in generating numbers and graphs.

The best use of discomfort surveys is probably for a single work area that has been targeted for improvements. The fact that changes are planned adds meaning to the survey and can increase the accuracy of the responses.

It is also possible to survey a whole workplace with the objectives of helping to prioritize areas or obtaining baseline information. Well done surveys can provide value, but they do entail a number of pitfalls. Accuracy can easily drop off if there is no perceived benefit to the survey participants. Furthermore, expectations can be raised beyond the ability of the organization to respond.

This type of survey is obviously subjective, and it is possible for results to be skewed because of other factors unrelated to the ergonomics purposes at hand. It is a tool, and like all tools, has its special uses, and may or may not be appropriate in any given situation.

These simple discomfort surveys differ from "active medical surveillance" where employees fill out similar forms, but include their names and typically more information about the type of discomfort experienced (sharp pain, throbbing pain, etc.). The purpose of active medical surveillance is to identify employees at earlier stages of problems in order permit speedier treatment. The questionnaires are usually completed in conjunction with a hands-on physical given by a medical provider, such as a nurse.

Tips

1. The first rule of employee surveys is, "Do not administer the survey if you are not prepared to respond to the results."
2. When several work areas are involved, it helps to provide codes for departments and jobs; otherwise, it is easy for people to use different names for the same thing, and thus confuse results.
3. It is helpful to administer the forms at the conclusion of a training session so that the purpose of the survey and its objectives can more clearly be understood (and so that all the forms can be collected as the participants leave the room).

There is often a concern that administering a discomfort survey can create problems. However, it has been the author's experience that nothing of the sort has occurred. If the purpose of the survey is make task improvements, and it is administered in an atmosphere of trust and sincerity, the experience has been positive.

22

Key Studies

This section provides summaries of studies that help provide the scientific basis of a number of the concepts described previously. Note that this is not a comprehensive review of all such studies, rather a selection of some of the more interesting ones that affect general workplace ergonomics.

Key studies

Each page provides some basic information:

1. Background of the study

2. Conclusions of research

3. Design implication

4. Cautions and comments about interpreting these results.

Please note there are may well be additional such conclusions and implications that are not listed here.

Occupational and Non-Occupational
Incidence of Carpal Tunnel Syndrome

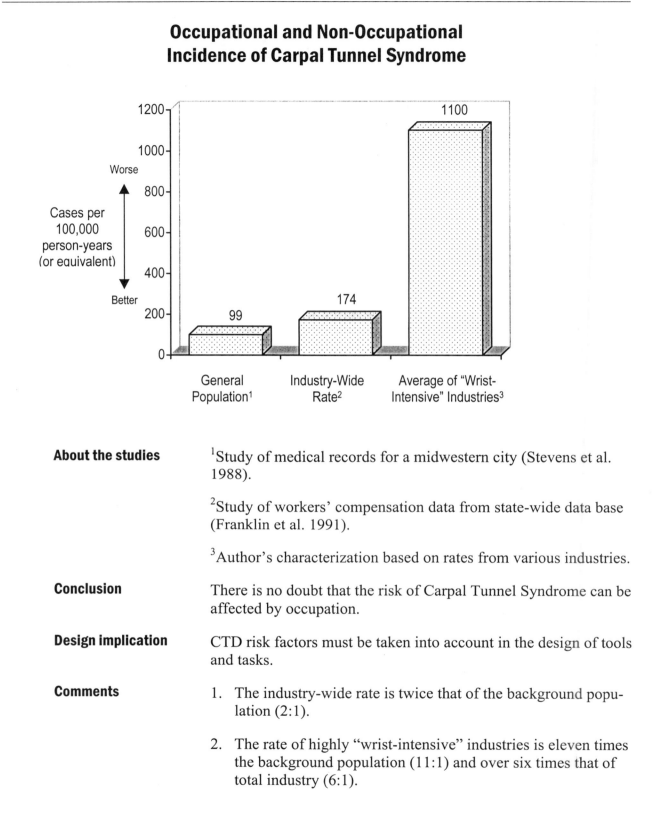

About the studies	[1]Study of medical records for a midwestern city (Stevens et al. 1988).
	[2]Study of workers' compensation data from state-wide data base (Franklin et al. 1991).
	[3]Author's characterization based on rates from various industries.
Conclusion	There is no doubt that the risk of Carpal Tunnel Syndrome can be affected by occupation.
Design implication	CTD risk factors must be taken into account in the design of tools and tasks.
Comments	1. The industry-wide rate is twice that of the background population (2:1).
	2. The rate of highly "wrist-intensive" industries is eleven times the background population (11:1) and over six times that of total industry (6:1).

Pressure Inside the Carpal Tunnel

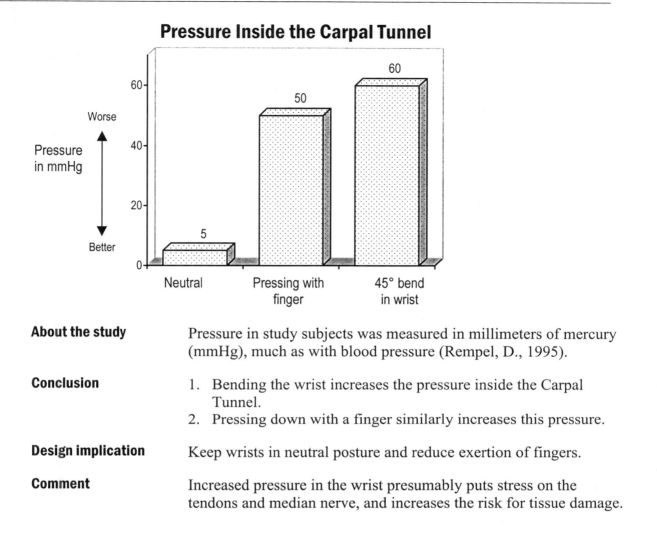

About the study   Pressure in study subjects was measured in millimeters of mercury (mmHg), much as with blood pressure (Rempel, D., 1995).

Conclusion
1. Bending the wrist increases the pressure inside the Carpal Tunnel.
2. Pressing down with a finger similarly increases this pressure.

Design implication   Keep wrists in neutral posture and reduce exertion of fingers.

Comment   Increased pressure in the wrist presumably puts stress on the tendons and median nerve, and increases the risk for tissue damage.

Professional Baseball Catchers with Left Hand Problems

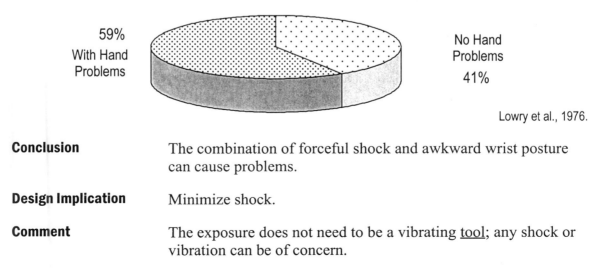

Lowry et al., 1976.

Conclusion   The combination of forceful shock and awkward wrist posture can cause problems.

Design Implication   Minimize shock.

Comment   The exposure does not need to be a vibrating tool; any shock or vibration can be of concern.

Static Work Postures and CTDs

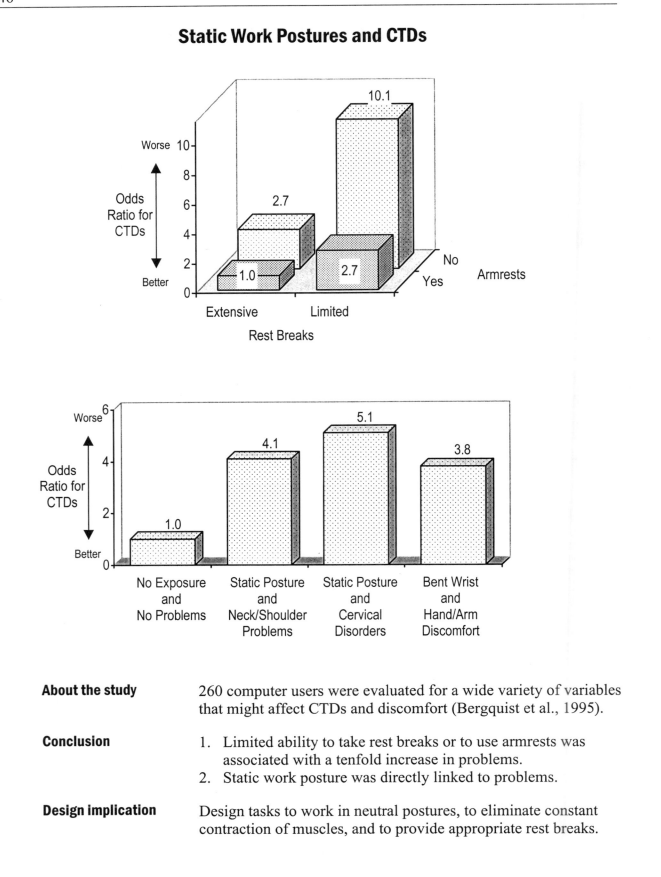

About the study	260 computer users were evaluated for a wide variety of variables that might affect CTDs and discomfort (Bergquist et al., 1995).
Conclusion	1. Limited ability to take rest breaks or to use armrests was associated with a tenfold increase in problems. 2. Static work posture was directly linked to problems.
Design implication	Design tasks to work in neutral postures, to eliminate constant contraction of muscles, and to provide appropriate rest breaks.

Wrist Posture and Strength

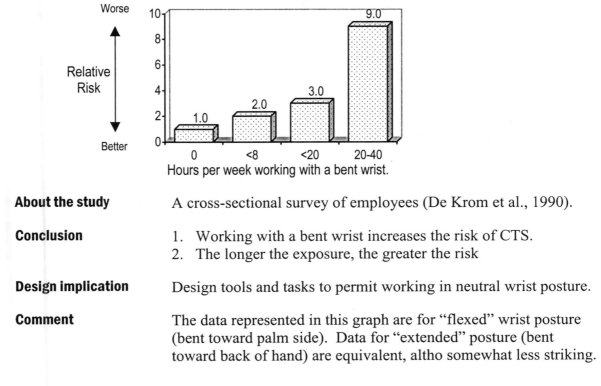

75% of full strength

100%; neutral posture; full strength

60% of full strength

45% of full strength

About the study	This is an relatively easy-to-do measurement; although one published reference is Putz-Anderson, 1988.
Conclusion	Bending the wrist reduces strength.
Design implication	Design to keep the wrist in its neutral posture.
Comments	Test this for yourself by using one hand to grasp tightly two fingers of the other hand, then pull out the two fingers. it much easier when the grasping hand is bent than when it is in the neutral posture.

Bent Wrist and Carpal Tunnel Syndrome

Worse

Relative Risk

Better

Hours per week working with a bent wrist.

0 → 1.0
<8 → 2.0
<20 → 3.0
20-40 → 9.0

About the study	A cross-sectional survey of employees (De Krom et al., 1990).
Conclusion	1. Working with a bent wrist increases the risk of CTS. 2. The longer the exposure, the greater the risk
Design implication	Design tools and tasks to permit working in neutral wrist posture.
Comment	The data represented in this graph are for "flexed" wrist posture (bent toward palm side). Data for "extended" posture (bent toward back of hand) are equivalent, altho somewhat less striking.

Repetition, Force, and Wrist CTDs

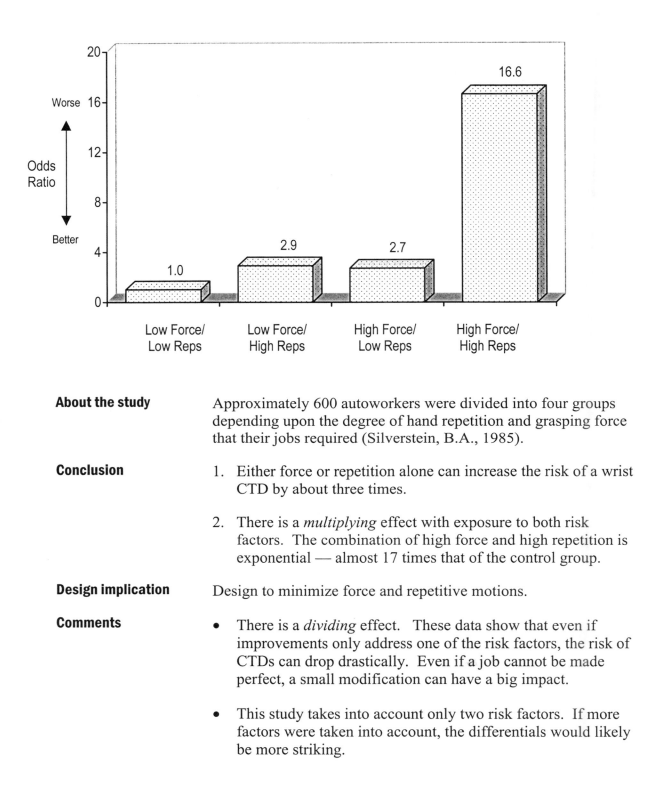

About the study

Approximately 600 autoworkers were divided into four groups depending upon the degree of hand repetition and grasping force that their jobs required (Silverstein, B.A., 1985).

Conclusion

1. Either force or repetition alone can increase the risk of a wrist CTD by about three times.

2. There is a *multiplying* effect with exposure to both risk factors. The combination of high force and high repetition is exponential — almost 17 times that of the control group.

Design implication

Design to minimize force and repetitive motions.

Comments

- There is a *dividing* effect. These data show that even if improvements only address one of the risk factors, the risk of CTDs can drop drastically. Even if a job cannot be made perfect, a small modification can have a big impact.

- This study takes into account only two risk factors. If more factors were taken into account, the differentials would likely be more striking.

Shoulder Problems and Height of Worksurface

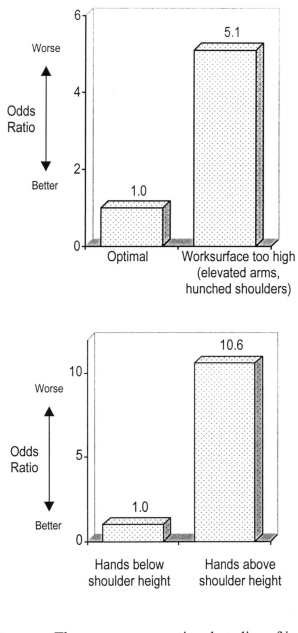

About the studies	These are cross-sectional studies of industrial workers (Hoekstra et al., 1994 and Bjelle, A. et al., 1979).
Conclusion	The obvious: working continuously with hunched shoulders or with raised arms increases risk of CTDs.
Design implication	• Use optimal heights to keep arms in the neutral posture (elbows at sides; shoulders relaxed).
	• Provide armrests or other support.

Shoulder Disorders, Repetition, and Static Load

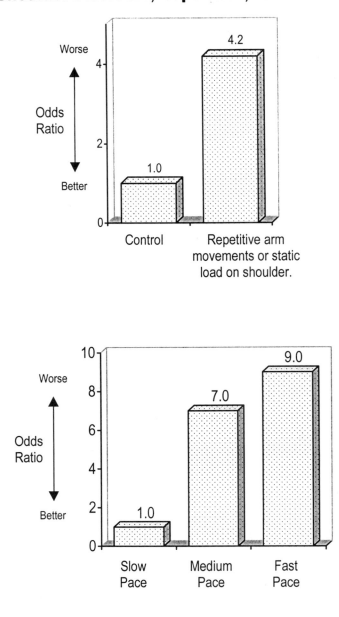

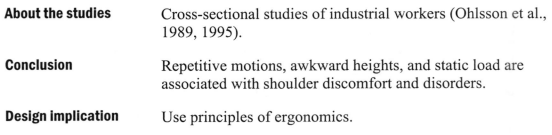

About the studies	Cross-sectional studies of industrial workers (Ohlsson et al., 1989, 1995).
Conclusion	Repetitive motions, awkward heights, and static load are associated with shoulder discomfort and disorders.
Design implication	Use principles of ergonomics.

Male/Female Strength Differences

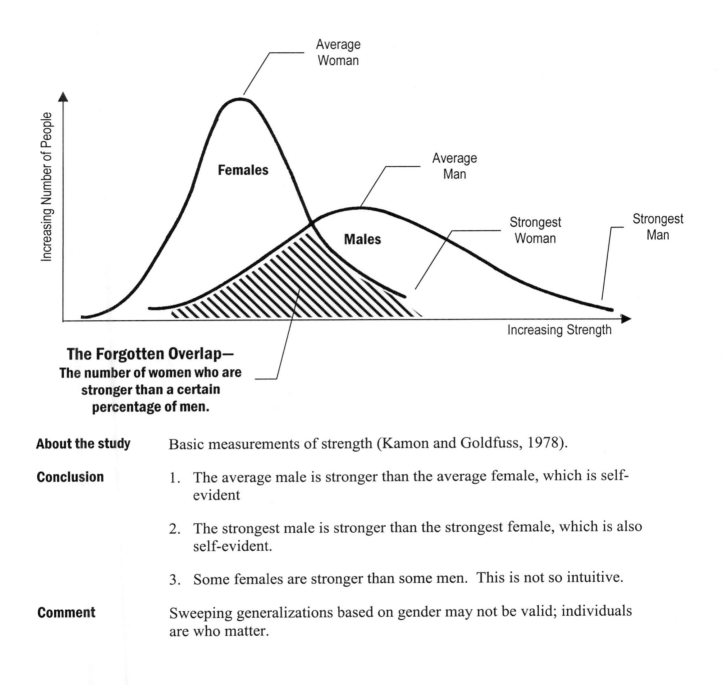

The Forgotten Overlap—
The number of women who are
stronger than a certain
percentage of men.

About the study	Basic measurements of strength (Kamon and Goldfuss, 1978).
Conclusion	1. The average male is stronger than the average female, which is self-evident
	2. The strongest male is stronger than the strongest female, which is also self-evident.
	3. Some females are stronger than some men. This is not so intuitive.
Comment	Sweeping generalizations based on gender may not be valid; individuals are who matter.

152

Maximum Duration Static Effort

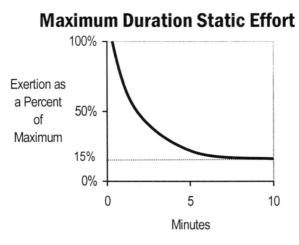

Roehmert, 1966, in Chaffin and Anderson, 1991

Conclusions

1. A muscle can be exerted at 100% of maximum for only a few moments. However, as the percentage of maximum exertion is reduced, it can be held for increasing lengths of time.
2. At about 15% of maximum effort, the curve levels, and thus can be held for a considerably longer period of time.

Design implications

When the load is less than about 15% of maximum, the exertion can be maintained for a longer period of time.

Comments

Other studies suggest limiting static exertion to less than 10% of maximum. This curve is often referred to as "Rohmert's Curve."

Elbow Posture and Shoulder Fatigue

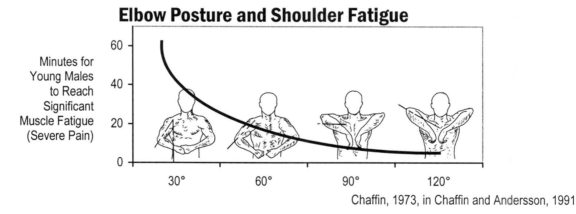

Chaffin, 1973, in Chaffin and Andersson, 1991

Conclusion

The expected time to reach significant muscle fatigue (severe pain) drops drastically with further extension of the elbows away from the body.

Design implication

Design tasks to keep the elbows at the side of the body.

Comment

This curve shows a relationship that is virtually self-evident, but all too often is not followed.

Arm Posture, Output, and Effort

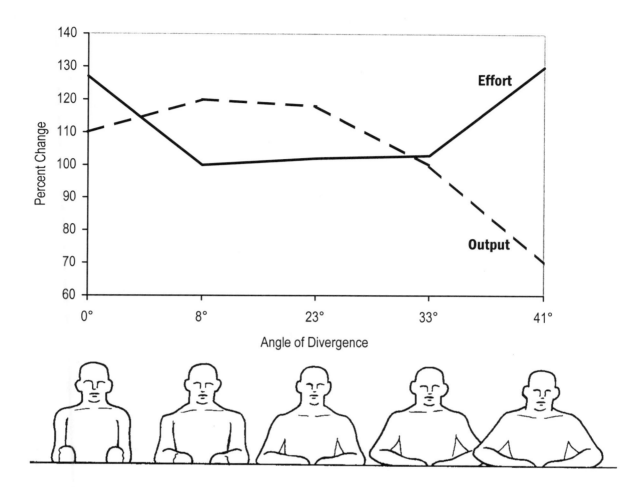

About the study A group of 12 grocery packers were evaluated for both performance and metabolic expenditure. The height of the work surface was manipulated to create the changes in arm posture (Tichauer, E.R., 1978).

Conclusion When the elbows are held either too close to the body or too far away, (a) performance decreases and (b) the effort required to do the job increases.

Design implication Work is best performed when the elbows remain in their neutral positions at the sides of the body.

Comment The conclusion merely confirms what seems intuitive. However, there is an additional implication that goes against our intuition. Traditionally, we have tended to think that the harder we work, the more we produce. This graph shows the opposite.

Here, the easier the work is designed to be, the greater the output.

Lower Back Compression and Various Postures

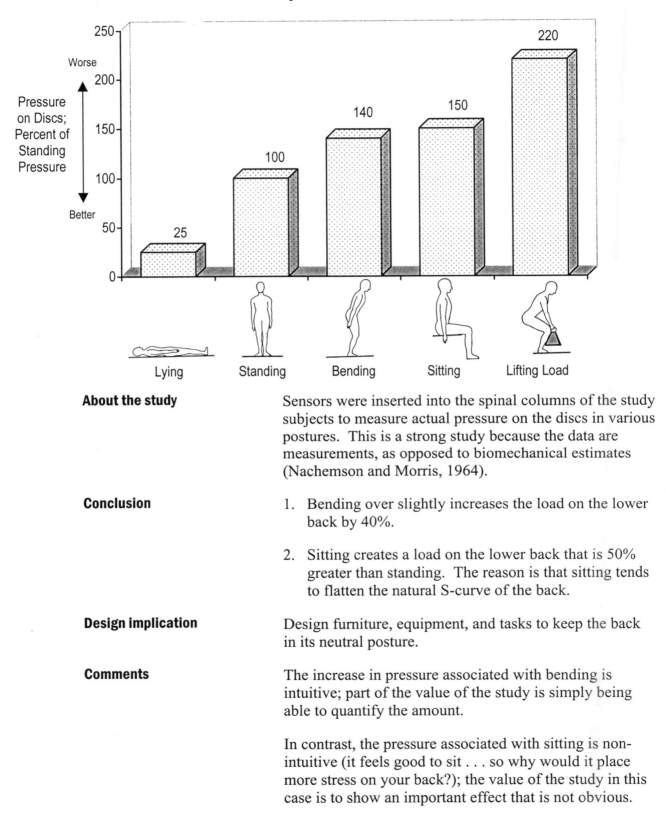

About the study

Sensors were inserted into the spinal columns of the study subjects to measure actual pressure on the discs in various postures. This is a strong study because the data are measurements, as opposed to biomechanical estimates (Nachemson and Morris, 1964).

Conclusion

1. Bending over slightly increases the load on the lower back by 40%.

2. Sitting creates a load on the lower back that is 50% greater than standing. The reason is that sitting tends to flatten the natural S-curve of the back.

Design implication

Design furniture, equipment, and tasks to keep the back in its neutral posture.

Comments

The increase in pressure associated with bending is intuitive; part of the value of the study is simply being able to quantify the amount.

In contrast, the pressure associated with sitting is non-intuitive (it feels good to sit . . . so why would it place more stress on your back?); the value of the study in this case is to show an important effect that is not obvious.

Lower Back Compression and Sitting Postures

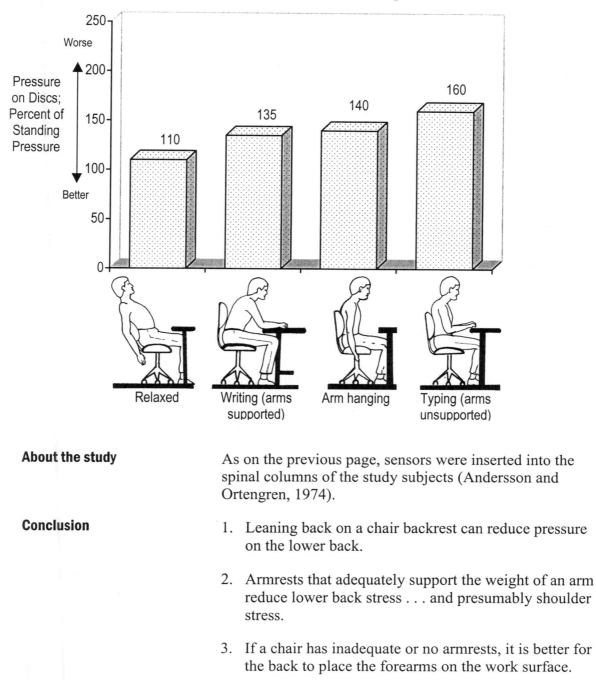

About the study

As on the previous page, sensors were inserted into the spinal columns of the study subjects (Andersson and Ortengren, 1974).

Conclusion

1. Leaning back on a chair backrest can reduce pressure on the lower back.

2. Armrests that adequately support the weight of an arm reduce lower back stress . . . and presumably shoulder stress.

3. If a chair has inadequate or no armrests, it is better for the back to place the forearms on the work surface.

Design implication

Chairs should have (a) armrests (which would need to be adjustable to accommodate varying heights of people), and (b) backrests that permit reclining.

Comments

Leaning back is different from slouching — *if* the natural curve of the lower back is maintained and supported, and *if* the weight of the upper torso is supported.

Benefits of Reclining and Lumbar Support

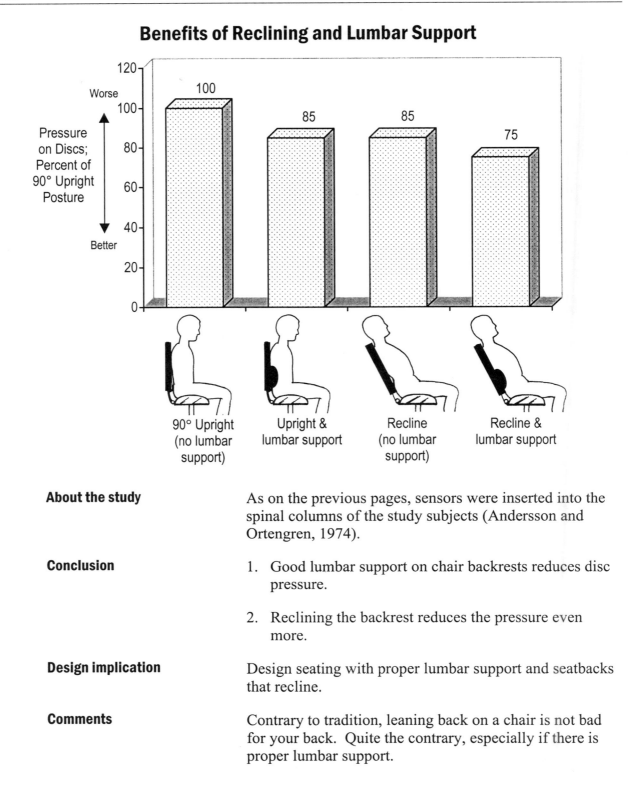

About the study	As on the previous pages, sensors were inserted into the spinal columns of the study subjects (Andersson and Ortengren, 1974).
Conclusion	1. Good lumbar support on chair backrests reduces disc pressure.
	2. Reclining the backrest reduces the pressure even more.
Design implication	Design seating with proper lumbar support and seatbacks that recline.
Comments	Contrary to tradition, leaning back on a chair is not bad for your back. Quite the contrary, especially if there is proper lumbar support.

X-Ray Evaluation of Various Postures

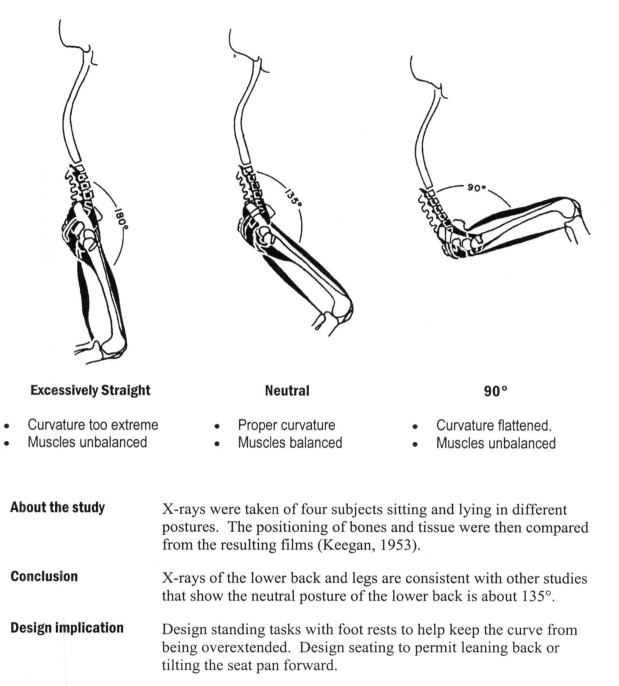

Excessively Straight	Neutral	90°
• Curvature too extreme • Muscles unbalanced	• Proper curvature • Muscles balanced	• Curvature flattened. • Muscles unbalanced

About the study X-rays were taken of four subjects sitting and lying in different postures. The positioning of bones and tissue were then compared from the resulting films (Keegan, 1953).

Conclusion X-rays of the lower back and legs are consistent with other studies that show the neutral posture of the lower back is about 135°.

Design implication Design standing tasks with foot rests to help keep the curve from being overextended. Design seating to permit leaning back or tilting the seat pan forward.

Comment This evaluation from a completely different perspective helps confirm the insights gained from the previous studies.

Physical Fitness and Back Injuries

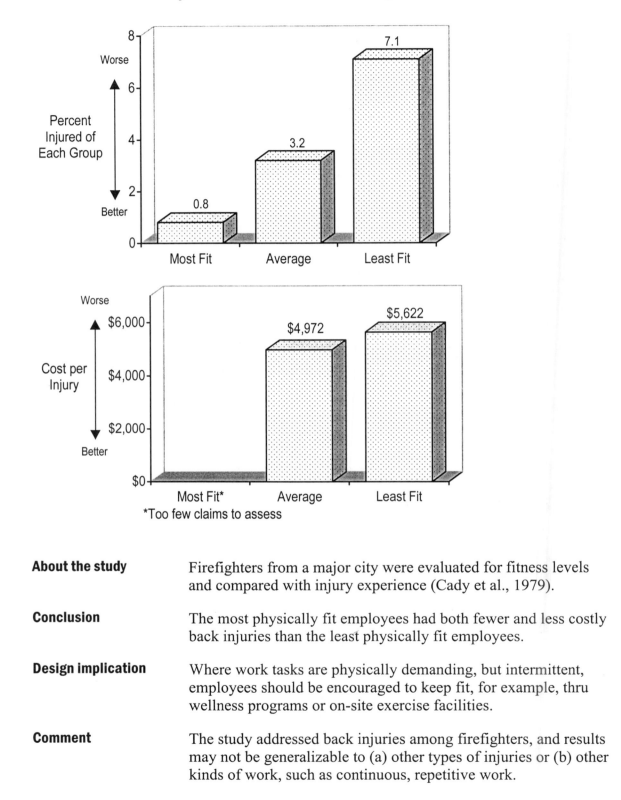

About the study	Firefighters from a major city were evaluated for fitness levels and compared with injury experience (Cady et al., 1979).
Conclusion	The most physically fit employees had both fewer and less costly back injuries than the least physically fit employees.
Design implication	Where work tasks are physically demanding, but intermittent, employees should be encouraged to keep fit, for example, thru wellness programs or on-site exercise facilities.
Comment	The study addressed back injuries among firefighters, and results may not be generalizable to (a) other types of injuries or (b) other kinds of work, such as continuous, repetitive work.

Breaks and Productivity

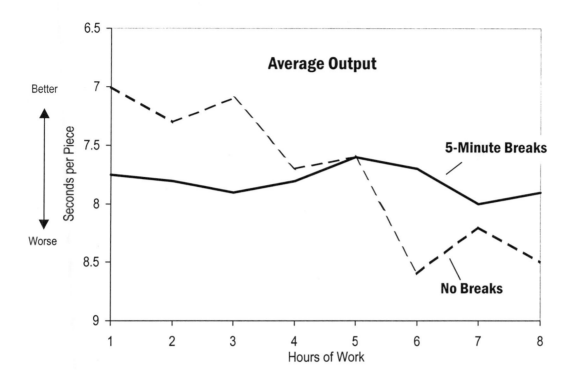

About the study	Various experiments were done with assemblers in the metalworking industry (Graf 1959, in Grandjean, 1997).
Conclusions:	1. Normal work patterns without hourly breaks result in steadily decreasing productivity.
	2. Five-minute breaks initially reduce the average output, but employees remain fresh and output remains steady.
Design implication	Build breaks into the work schedule.
Comments	This information is not exactly new. There are many studies of this sort dating back to the 1920s to 1950s. The early time and motion experts seem to have recognized the importance of preventing fatigue much more than is emphasized today.

Work Organization — Stress Levels

Repetitiveness

Adrenaline Levels

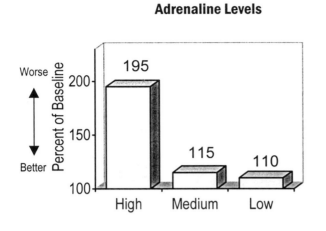

Feelings of "Irritation"

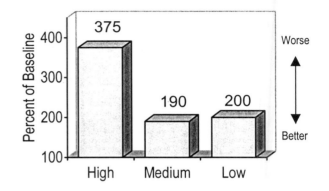

Task Variety

Adrenaline Levels

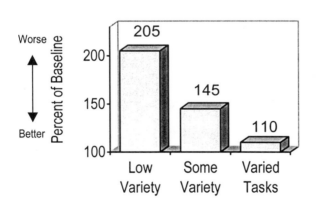

"Relaxed Feeling"

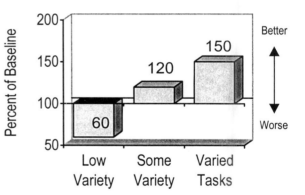

Constrained Posture

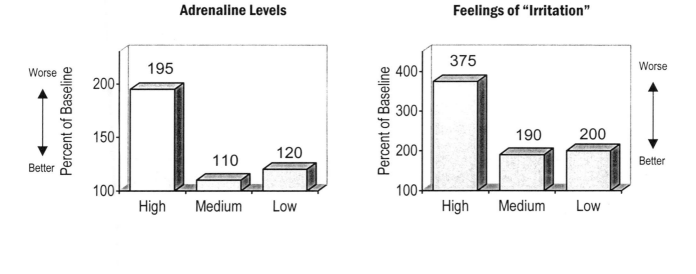

Control of Work Pace

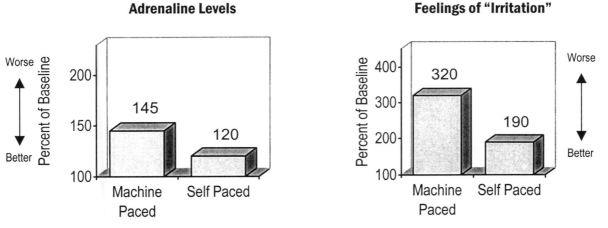

Frankenhäuser, 1981

Conclusion	Various factors of work organization can trigger the physiological stress response.
Design Implication	Where possible, jobs should be designed to have (a) low repetitiveness, (b) task variety, (c) permit neutral postures and movement, and (d) be self-paced.
Comment	Note that the objective measurements of adrenaline generally match the subjective statements of feelings.

Work Organization — Overtime and Stress

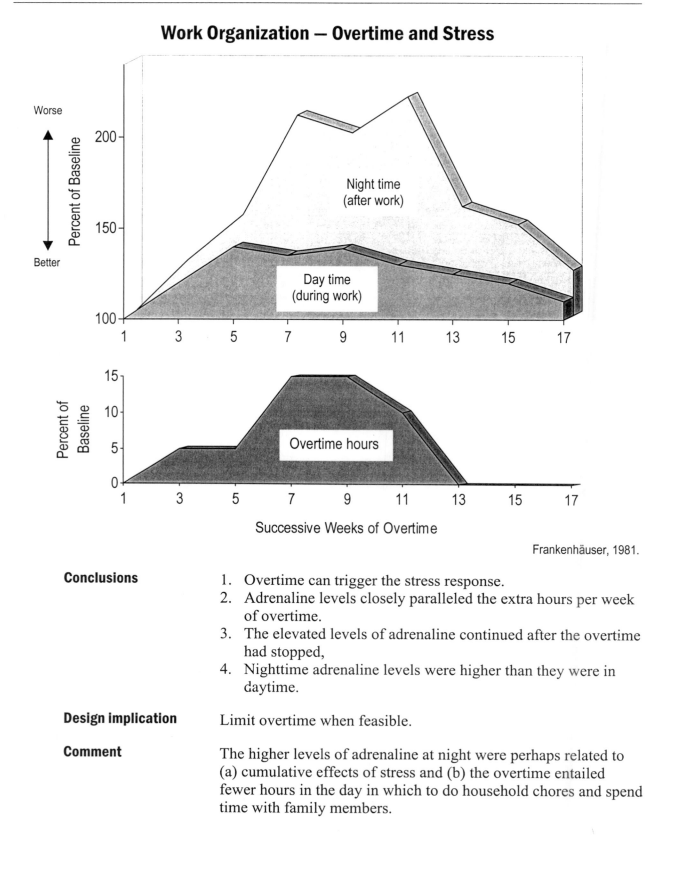

Frankenhäuser, 1981.

Conclusions	
	1. Overtime can trigger the stress response.
	2. Adrenaline levels closely paralleled the extra hours per week of overtime.
	3. The elevated levels of adrenaline continued after the overtime had stopped,
	4. Nighttime adrenaline levels were higher than they were in daytime.

Design implication   Limit overtime when feasible.

Comment   The higher levels of adrenaline at night were perhaps related to (a) cumulative effects of stress and (b) the overtime entailed fewer hours in the day in which to do household chores and spend time with family members.

Stressful Conditions and CTDs

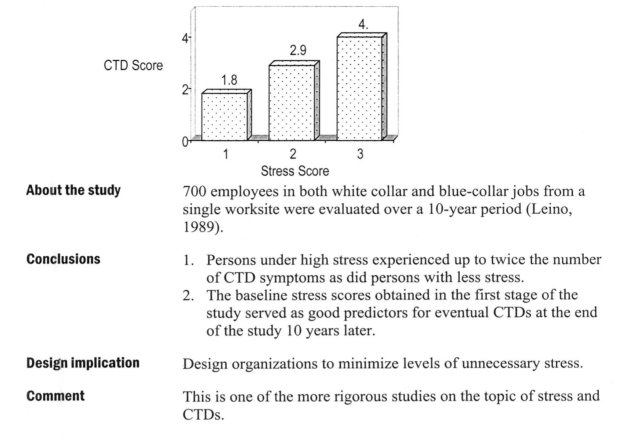

About the study	700 employees in both white collar and blue-collar jobs from a single worksite were evaluated over a 10-year period (Leino, 1989).
Conclusions	1. Persons under high stress experienced up to twice the number of CTD symptoms as did persons with less stress. 2. The baseline stress scores obtained in the first stage of the study served as good predictors for eventual CTDs at the end of the study 10 years later.
Design implication	Design organizations to minimize levels of unnecessary stress.
Comment	This is one of the more rigorous studies on the topic of stress and CTDs.

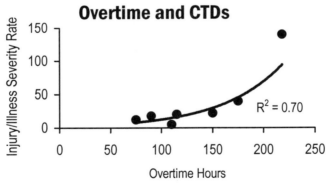

About the study	Data on overtime hours and on CTDs were obtained from the seven major production facilities of a major corporation. Each plant had about 1000 employees and produced the same product using equipment and methods that were nearly identical (author's client data).
Conclusions	Increased hours of overtime are related to CTDs.
Design implication	The obvious: Avoid excessive overtime.
Comment	Additional, more rigorous studies are needed.

Overtime and Lost Productivity

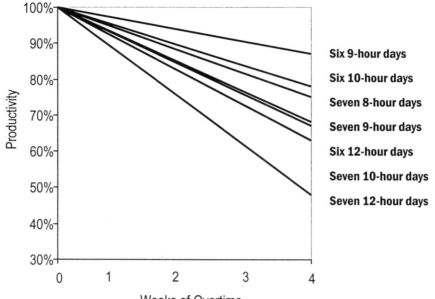

Schedule	Total Hours Per Week	Effective Hours After 4 Weeks	Productivity After 4 Weeks*
Six 9-hour days	54	47	87%
Six 10-hour days	60	47	78%
Seven 8-hour days	56	42	75%
Seven 9-hour days	63	43	68%
Six 12-hour days	72	48	67%
Seven 10-hour days	70	44	63%
Seven 12-hour days	84	40	48%

Percentage of initial productivity.

National Electrical Contractors Association, 1969

Conclusions

1. Overtime results in significant loss of productivity.

2. Loss of productivity is diffused, not concentrated at the end.

3. Loss of productivity is progressive with successive weeks.

4. Overtime schedule affects productivity; that is, seven 8-hour days are less productive than six 9-hour days.

5. Productivity losses are related to errors, absenteeism, and accidents in addition to reduced work pace.

Design implication

The obvious: Avoid excessive overtime.

23

Problem-Solving Process

An initial focused program is often necessary to set in motion an on-going process. Several basic elements of such a program are outlined below, which every employer should adopt in one form or another. Despite differences between type of industry and size of business, the following framework provides an approach for integrating ergonomics into day-to-day worklife. These program elements are described in full detail elsewhere (MacLeod, 1998, 1999) and are addressed here only as they apply to engineering aspects of the process.

Program Elements

1. **Organization** — a plan for getting organized, assigning responsibility, and involving people.

2. **Training** — an effort to provide training in ergonomics to personnel at all levels of the organization.

3. **Communication** — systems for communicating activities and progress.

4. **Task Analysis** — a systematic way to review all work areas for needed improvements.

5. **Making Improvements** — the key part of the process; making improvements whenever feasible.

6. **Medical Management** — procedures and protocols for identifying and treating employees with symptoms of CTDs.

7. **Monitoring Progress** — ways to measure and evaluate the program.

There are a variety of ways to implement ergonomics programs, all depending upon the needs and resources of each facility.

Individual companies vary in size, complexity of operations, and style of management. The level of activity and resources committed to the program depend upon the extent of existing problems. In small organizations, these program elements can be achieved with relatively modest efforts. In large organizations, or ones with severe problems, considerable work may be necessary to implement a good program. In short, there is no one best way to set up a program, as long as the basic program elements are implemented in an appropriate fashion.

Standards

There are a number of standards and guidelines that have been developed in recent years that affect ergonomics programs:

- California OSHA's *Prevention of CTDs* standard.
- Federal OSHA's *Ergonomics Program Management Guides for Meatpacking Plants*
- ANSI Z-365 *Control of CTDs* standard
- Federal OSHA's proposed *Ergonomics Program Standard*
- OSHA 5(a)(1) enforcement practice, administrative law rulings

With few exceptions, most of these documents do not address specific issues like the height of workstations or shapes of hand tools. Rather, they encourage management to set up programs to integrate ergonomics into everyday business life. Although these vary in their detail, on the whole they amount to the set of core requirements itemized on the previous page.

Tips and Ideas

1. Organization

Overall considerations for establishing an ergonomics process include management commitment, a written program, and formalizing the assignment of responsibility. Items that affect engineering aspects of the process include the following:

Team approach

Ergonomics works well using a team process, both for overall site-wide decision-making and for analyzing specific work areas. The composition of a team can vary widely with each workplace. A "team" can sometimes simply be a few people talking informally at the workstation. Other times, the team is more formal, including a cross-section of people:

- management, including top and first line
- production employees and union representatives
- maintenance and engineering
- health care providers
- equipment vendors
- outside experts, such as professional ergonomists

In particular, as specific tasks are reviewed and improvements sought, it is helpful to have participation of employees. The advantages to this approach are several. A fuller picture of the issues often is gained because of input from people with various perspectives. Consequently, there can be a higher quality of understanding of the problem and potential improvements. Furthermore, as a result of participation, there is often an increased acceptance of the proposed changes, making it easier to implement these changes. Finally, because of the early-on involvement of people, it is often easier to refine the changes and make follow-up workstation modifications.

Consult with users

Tool and workstation design often differs from other aspects of facility planning in an important way. It is quite possible, for example, to have an electrician review the needs of the workplace and design a wiring system that performs perfectly adequately without talking to any users. Setting up a workstation is usually different in that there are often too many nuances in the operation that are only known to people who have actually done the work. Consequently, for the best designs, communication between users and designers is needed.

Employee involvement in general has become an effective approach towards improving operations from many perspectives. Many companies have achieved a variety of paybacks because of involving employees in business operations, including improved morale, more informed decision-making, and innovation. As a final comment, if a company has never had a formal way to involve employees, a workplace ergonomics program is an excellent place to start.

2. Training

Proper training for personnel at all levels of the organization provides the foundation on which the process is built. Everyone needs to know the rules of work.

Aspects of training related to engineering generally focus on the basic principles of principles, strategies for making improvements, and quantitative methods. Employees and supervisors need a good understanding of these principles in order to contribute to the evaluation of tasks and suggest possible improvements.

3. Communication

A good system of communications is crucial to the workplace ergonomics process. If people are not kept updated about plans and status of projects, they may assume that nothing is happening and the entire process can be jeopardized. Aspects of communication that are particularly important for engineering efforts include the following actions:

Notify affected employees in advance when certain areas are going to be modified.

Solicit input on improvements needed in work areas.

Provide feedback to personnel who have provided ideas for improvement on the status of those suggestions.

Explain why videotapes are being taken of jobs, and why questions are being asked about aches and pains.

Maintain a list of improvements that have been made. This list provides the basis for other communications.

4. Task Analysis

Developing systematic approaches to evaluating all tasks in a workplace is a vital element of engineering efforts.

Ergonomic job analysis has been the source of confusion, especially when recommended or required as part of regulatory action. Textbooks and scientific papers often inadvertently give the impression that job analysis is complex and involves elaborate equipment.

The tools of job analysis are varied: Some are basic and some complex. The technique and approach used must fit the needs and goals of the specific workplace. Unfortunately, too many job analyses are being generated today that are far too long and academic for the need at hand. It is like giving a brain scan to a patient with a head cold, when a simple prescription of aspirin and chicken soup would do.

A good job analysis can be quite simple and yet provide good insight into issues and potential improvements. The key is to be systematic, have a good understanding of ergonomic principles and be clear about which of several goals is the aim of the analysis.

Documenting Problems Versus Finding Solutions

A primary source of error is not differentiating between documenting a problem versus finding a solution. If the goal is to document problems, then rigor and precision in conducting the analysis is critical. However, if the goal is to find improvements, the step of determining the problem is often fairly simple. The ergonomic issues might be fairly obvious, and there is no particular need to stop to evaluate the problem any further. One can quickly recognize the problems, then almost immediately focus on options for improvements.

This distinction is important, since if the goal is to improve things, then spending time continuing to document the problem can be a distraction and counter-productive. Further study of the problem may lend no insights to problem-solving at all.

Analysis versus Measurements

A second, related source of error is to confuse "analysis" with "measurements." *Analysis* means to separate a whole into its constituent parts. In the context of preventing CTDs, job analysis involves breaking down a task into its elemental steps and identifying CTD risk factors for each separate part of the body (neck, back, wrists, etc.). Measuring risk factors and formally characterizing jobs, even tho it is important in its place, may not be necessary. The task analysis can be done using a checklist or indeed without using pen and paper at all.

Cost-Saving Distinctions

If task analysis is approached in the wrong way, it could end up costing more than it is worth. It is crucial to be clear on goals, because the choice of the type of task analysis to be used can vary.

In most workplaces, making improvements is the primary goal, in which case a low-tech, common-sense approach is often sufficient. However, when more rigor is needed, the measurement tools described in this book can provide great value.

Checklists and Worksheets

Well-designed checklists and worksheets can serve as simple, yet effective, tools. Supervisors and employee representatives can easily be trained to identify issues using a good checklist with virtually the same accuracy as a professional ergonomist. A good checklist can serve as a "mind-jogger" of risk factors and can help the evaluator to be systematic and to break the task down into specific elements (see MacLeod, 1999).

Videotapes

The video camera is the prime working tool of the workplace ergonomist. Videos help focus attention, so it is possible to notice issues on the video that normally not be observed live in the workplace. Videos can be played back multiple times as well as in slow motion to facilitate viewing a specific step. Finally, videos can be shown in a conference room where it is sufficiently quiet to have a discussion and where a team can be gathered for review, including the employees performing the task on the video. These activities are usually not possible on the workplace floor.

Task Analysis

Analysis literally means to take a whole and break it into its sub-parts (i.e., the opposite of *synthesis*). Tasks are often too complex to gain much insight by looking at them as a whole. However, by focusing on one item at a time, it is possible to learn a great deal.

In this context, task analysis means to break a job down into:

- Steps of the job — "reach for part A, insert into part B, etc."
- Various parts of the body — wrists, arms, and back.
- Various ergonomics issues — awkward postures, force, and so forth.

Then, for each ergonomics issue affecting a specific body part during a particular step of the job, one asks, "Is there an easier way to do this by applying the basic principles of ergonomics?" Brainstorming is often extremely helpful, as are other techniques of group problem solving.

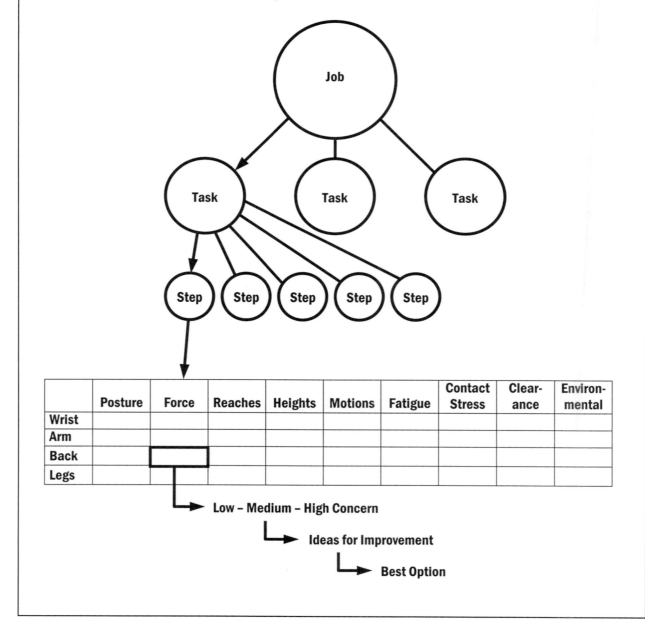

	Posture	Force	Reaches	Heights	Motions	Fatigue	Contact Stress	Clear-ance	Environ-mental
Wrist									
Arm									
Back									
Legs									

Low – Medium – High Concern

Ideas for Improvement

Best Option

The Problem-Solving Process

1. Identify priority tasks

Where have most sprains and strains type injuries occurred?

What tasks are the most strenuous or frustrating?

What are the tasks that people dislike doing the most?

What tasks occur most frequently?

What are most crucial to accomplish quickly?

What are most crucial to accomplish correctly?

What tasks would be most easy to fix?

2. Fact-finding

Evaluate the task using an ergonomics worksheet

Discuss issues with employees and supervisor

Videotape the task

Identify the important ergonomic issues

3. Brainstorming

While observing the task, think of creative ways to make improvements

As a team, meet in a conference room, watch the video, review the worksheet, and brainstorm possible ideas

4. Develop Action Plans

What ideas are feasible? Are there any quick fixes?

What additional information is needed? Who will do this?

What types of long-term changes are possible?

5. Evaluate Results

Discuss with employees

Measure effects, as appropriate

Apply continuous improvement

5. Making Improvements

Identifying feasible improvements is the key part of the whole process. All other activities are subordinate to this goal. Improvements can be found through both standard vendor sources and traditional in-house engineering.

Short-range improvements

After initial job evaluations are made, many small and inexpensive improvements can often be identified and sometimes be completed literally overnight. Typical examples are layout changes to improve heights and reaches, the purchase of anti-fatigue mats and chairs, or changing work methods.

Long-range equipment development

Long range improvements are sometimes also needed. This may involve some research and development before a change becomes technically feasible. In many cases, these changes may dovetail with long-range plans for quality and productivity improvement.

Trial and error

Whether short or long range, everyone should recognize that ergonomics improvements are not always straightforward and that a period of experimentation and trial and error is often needed to find a good modification. The concept of "continuous improvement" is important — a job improvement is planned, then implemented, then evaluated, and then refined in an ongoing process.

Creative thinking

Many times the problems and solutions are apparent and the action is a pro forma effort of ordering equipment from vendors or writing a standard work order. On the other hand, there can be countless options for making improvements. For example, before purchasing a lift table to reduce the risk of injury from lifting, it can be worthwhile to determine if there are different ways altogether of accomplishing the same task. It might be possible to identify a completely new and better way of doing the job.

Continuous Improvement

People too often adopt a black-and-white stance regarding problems and solutions. The phrases one hears is "ergonomically correct."

In reality, "ergonomic" is relative. For the most part, the word *improvement* is a more suitable term than *solution*, since once a task is made better, it can almost always be improved upon at some point.

Thus, "ergonomically correct" depends to a large degree on what one compares it with. The whole process should be viewed from the perspective of continuous improvement.

Multiple Options

Furthermore, the phrase "options for improvement" is a good one to keep in mind when trying to address ergonomics issues. People also often presume that there is one — and only one — solution to problem. Experience, however, shows that there are typically multiple options for making improvements. After a period of brainstorming and evaluation, some of these options may well show more promise, but the process should at least begin with an open mind to multiple options. In the end there may still be more than one viable solution.

Probing questions

What improved types of tools are possible?

Are fixtures or hold-down devices possible?

What types of mechanical assists might be used?

What changes in layout would help?

Would improvements in the overall material handling system help?

Would changes in the overall work process help?

Is there a completely different way of doing the job?

Mindset questions

If you were a Yankee inventor living in 1820 and had no electricity or power, how would you do this job?

If you had unlimited resources, what would you do?

If there is an automatic way of doing this task, but too expensive or not feasible in this case, is there some half-way modification that is feasible?

Have you ever seen a different way of doing this task? What implications does that have in this case?

Does a similar task exist in another industry? How do they do it there? What are the implications for you?

Sources of Ideas

In-house capabilities

Once the basic concepts of ergonomics are understood and problems identified, almost anyone can have an idea for improvement. In-house engineers, managers and production employees can be the source of many improvements once pointed in the right direction.

Literature

An increasing amount of information is becoming available on improvements:

- Industry trade magazines, including magazines from industries *other* than one's own.

- Publications on ergonomics

- Vendor catalogs — a wealth of ideas can be found in these catalogs, especially if you are creative and can think of unconventional uses of standard equipment.

Pitfalls in purchasing ergonomic products

- "Ergonomic" products that are used inappropriately are not ergonomic.

- Any product that leads to improvements can be "ergonomic."

- There is no certification or testing procedure to determine if a product is truly ergonomic . . . buyer beware!

To be able to evaluate products, one must understand the underlying principles of ergonomics.

Ergonomics consultants

Ergonomists can be a useful part of this process, by challenging traditional thinking and by providing new insights into the design concepts. The ergonomist can provide a new set of eyes to see workplace issues to which in-house staff may have become accustomed. Finally, the ergonomist may have experience with improvements in other industries that may have value.

Contractors and equipment suppliers

A good idea is to invite contractors and equipment suppliers to meetings and training sessions. Especially, those companies that rely on local contractors to build and install equipment. Make sure that they know of your efforts and that you expect equipment that has undergone ergonomic review.

Other Facilities

Additional sources of ideas can be found by visiting other facilities, both in your industry and other industries. Conferences and trade shows can also be valuable.

Tracking system

As part of this process, a good filing system may be needed to keep track of results of job evaluations, ideas for improvement, planned changes, and overall progress. The system can range from a simple notebook or file drawer to a computer spreadsheet or data base system.

Improve continuously

One of the tenets of quality is that there is really no end to the change process. Improvements must continually be made. No one should have the expectation that an ergonomics program will be put in place and at some point be over and done with. Continuous improvement is the name of the game.

6. Medical Management

Most of the aspects of medical management of CTDs are outside the scope of this book, altho essential in their own right. The most important activity relevant here is the system of restricted duty; by using the tools described in this book it is possible to document physical task requirements to help identify proper jobs for employees with physical restrictions.

7. Monitoring Progress

The ergonomics program should be evaluated periodically. Several approaches can be used, but the ones most important in this context are:

Ergonomics log

It is helpful to maintain a log of all improvements made that have had an ergonomic impact. This log may include items such as department, job, ergonomics issue involved, improvement, cost, date, and contact person.

Keeping this log provides value in the following ways:

- The log reminds everyone of accomplishments, thus keeping momentum going and the morale of the team high. Without such a list, it is easy to forget the things completed on a daily basis and there can be an unwarranted feeling of ineffectiveness.

- The log provides a basis for communicating results to both managers and employees. It can be impressive to report that "in the past year at the XYZ Company there have been 326 improvements in the workplace as a result of the ergonomics effort."

Special studies

A variety of factors can be evaluated to measure the effect of the ergonomics program:
- "Before-and-after" comparisons of tasks
- Employee survey results
- Workers' compensation costs
- Turnover and absenteeism
- Quality and productivity

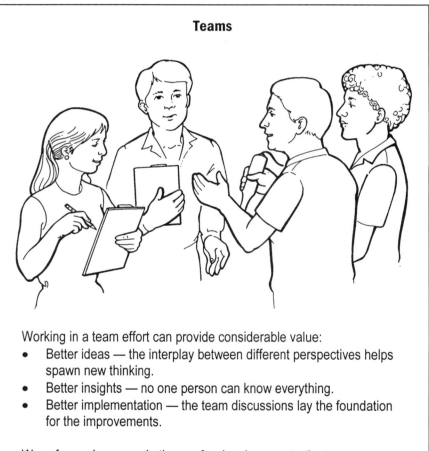

Teams

Working in a team effort can provide considerable value:
- Better ideas — the interplay between different perspectives helps spawn new thinking.
- Better insights — no one person can know everything.
- Better implementation — the team discussions lay the foundation for the improvements.

Ways for engineers and other professionals promote the team process include:
- Solicit input continuously.
- Work the production jobs from time to time to gain a better understanding of the task demands.
- Get to know the production employees as individuals.
- Help create an atmosphere of innovation.

Videotaping for Ergonomics Task Analysis

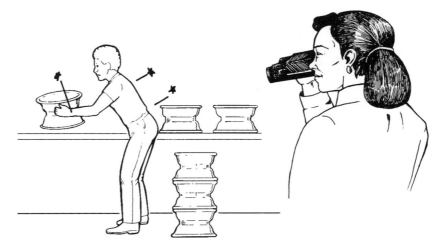

The video camera is the primary tool of the workplace ergonomist.

1. Before starting to tape be sure that all settings on the video camera are correct. It is usually a good idea to (a) check the battery power, and (b) turn on the clock or time code function.

2. It is a good idea to record the name of the job at the beginning of each separate job or task taped. This information can be written on a sheet of paper, then taped for several seconds.

3. If the people in the area are not familiar with you or why you are videotaping them, it is a good idea to introduce yourself and explain your objectives.

4. Tape all aspects of the task. This usually includes at least 5-10 minutes of video and is best if at least 10 cycles of the task are included. When in doubt, tape more than needed, rather than less, since it is always possible to fast forward.

5. Hold the camera still. For hand-held shots, it is sometimes helpful to place your right elbow firmly against your side and use your left hand under the camcorder to support it. In addition, place your eye firmly against the viewfinder cup. Do not walk with the camera unless absolutely necessary to record the task. If you must move or "pan" the camera, do so slowly to reduce recorded camera movement.

6. Begin taping with a whole body shot of the worker to provide perspective (be sure to include the chair or surface on which the employee is standing) and hold this for about 2-3 cycles. Then stop, find another angle, zoom in on the hands and arms, or other part of the body at issue, and begin taping again.

7. Continue using various angles and zooms until all aspects the task are documented. It is best to choose angles that permit subsequent evaluation of various shoulder, arm, hand, neck, back, and leg postures. Tape from both sides, in front, and overhead if possible.

8. It is helpful to video the task at the same time as the team conducts its live evaluation.

9. Keep completed tapes in an archive, rather than tape over previously filmed tasks. Tape is inexpensive, and it is helpful to refer to old footage at some point.

Websites

The web is becoming an increasingly good source of up-to-date information. The following are some useful sites, from which one can link to many more. One can also search for "ergonomics," "cumulative trauma disorders," and other similar phrases.

www.mcmaster.com — this industrial supply company constitutes the largest single source for ergonomics products of all types.

www.grainger.com — another excellent industrial supply company with hundreds of good products.

www.thomasregister.com — a master listing of American companies and their products.

www.mhesource.com — a commercial site that contains a list of vendors of material handling equipment.

www.ergoweb.com — a commercial site that contains background information and vendor information, as well as access to software-based task analysis tools.

www.nexgenergo.com — a vendor for software-based task analysis tools.

http://cseriac.flight.wpafb.af.mil — a site that provides anthropometric data based on U.S. Air Force personnel, plus other helpful tools and materials.

www.alimed.com — a vendor of ergonomics products.

www.cdc.gov/niosh/ergopage.html — the official site of the National Institute for Occupational Safety and Health (NIOSH), which contains an extensive review of scientific literature.

www.osha.gov — the official site of the Occupational Safety and Health Administration (OSHA), which has a page and links devoted to ergonomics, including updates on development of CTD prevention standards and guidelines.

www.clmi-training.com — a source for video-based ergonomics training programs for supervisors and employees.

www.danmacleod.com — the author's personal site, which contains a variety of supplementary material.

www.taylorandfrancis.com — the publisher's site, which contains references to other ergonomics publications.

References

Part I — The Rules

Chaffin, D.B. and Andersson, G.B.J., 1991. *Occupational Biomechanics*, Second Edition. New York: John Wiley and Sons.

Grandjean, E., 1997. *Fitting the Task to the Man*, 4th Edition. New York: Taylor and Francis, Ltd.

Kodak, 1983. Human Factors Section, Eastman Kodak Company. *Ergonomic Design for People at Work. Volumes I and II.* New York: John Wiley & Sons.

Kroemer, K.H.E., Kroemer, H.J., and Kroemer-Elbert, K.E., 1990. *Engineering Physiology*. 2nd Edition. New York: Van Nostrand Reinhold.

McCormick, E. and Sanders, M., 1989. *Human Factors in Engineering and Design*, 6th Edition. New York: McGraw Hill.

Pheasant, S., 1996. *Bodyspace: Anthropometry, Ergonomics and the Design of Work*, Second Edition. New York: Taylor and Francis.

Putz-Anderson, V., 1988. *Cumulative Trauma Disorders: A Manual for Musculoskeletal Disorders of the Upper Limb.* New York: Taylor and Francis.

Ranney, D., 1997. *Chronic Musculoskeletal Injuries in the Workplace.* Philadelphia: W.B. Saunders Company.

Tichauer, E.R., 1978. *The Biomechanical Basis of Ergonomics.* New York: John Wiley and Sons.

Part II — Measurements and Guidelines

Quantitative Methods

American National Standards Institute, ANSI Z-365 Committee, 1995. *Control of Work-Related Cumulative Trauma Disorders,* Draft Appendix. Itasca, IL: National Safety Council.

Bloswick, D., 1999. *Estimate of Back Compressive Force,* self-published training materials. University of Utah, Salt Lake City: bloswick@eng.utah.edu

Pheasant, S., 1996. *Bodyspace: Anthropometry, Ergonomics and the Design of Work*, Second Edition. New York: Taylor and Francis.

Chaffin, D.B. and Andersson, G.B.J., 1991. *Occupational Biomechanics*, Second Edition. New York: John Wiley and Sons.

Mathiowetz V., Kashman N., Volland G., Weber K., Dowe M. and Rogers S., 1985. Grip and Pinch Strength: Normative Data for Adults, *Arch. Phys. Med. Rehabil.* (Vol. 66).

Mital, A., Nicholson, A.S., and Ayoub, M.M., 1993. *A Guide to Manual Materials Handling.* London: Taylor and Francis.

NIOSH, 1994. *Applications Manual for the Revised Lifting Equation.* Department of Health and Human Services, Center for Disease Control, National Institute for Occupational Safety and Health. DHHS (NIOSH) Publication No. 94–110. Springfield, VA: U.S. Department of Commerce Technology Administration, National Technical Information Service.

Snook S. and Ciriello, V., 1990. The design of manual handling tasks: Revised tables of maximum acceptable weights and forces. *Ergonomics* 34(9):1197-1213.

Key Studies

Andersson, G.B.J. and Ortengren, R., 1974. Myoelectric back muscle activity during sitting, *Scand. J. Rehab. Med.* 3, 104–114, & Suppl. 3, 73–133.

Bergkvist, U., Wolgast, E., Nilsson, B., and Voss, M., 1995. Musculoskeletal disorders among visual display terminal workers: individual, ergonomic, and work organization factors. *Ergonomics* 38:763–776.

Bernard, B., Sauter, S., Fine, L.J., Petersen, M., and Hales, T., 1994. Job task and psychosocial risk factors for work-related musculoskeletal disorders among newspaper employees. *Scand. J. Work Environ. Health* 20(6):417– 426.

Bjelle, A., Hagberg, M., and Michaelsson, G., 1979. Clinical and ergonomic factors in prolonged shoulder pain among industrial workers. *Scand. J. Work Environ. Health* 5(3):205–210.

Cady, L.D., Bischoff, D.P., O'Connell, E.R., Thomas, P.C., Allan, J.H., 1979. Strength and fitness and subsequent back injuries in firefighters. *J. Occ. Med.* 21(4).

Chaffin, D.B. and Andersson, G.B.J., 1991. *Occupational Biomechanics*, Second Edition. New York: John Wiley and Sons.

deKrom, M.C.T., Kester, A.D.M., Knipschild, P.G., and Spaans, F., 1990. Risk factors for carpal tunnel syndrome. *Am. J. Epidemiol.* 132(6):1102–1110.

Frankenhäuser, M., 1981. Coping with stress at work. *Int. J. Health Services.* 11(4):491-510.

Franklin, G.M., Haug, J., Heyer, N., Checkoway, H., and Peck, N., 1991. Occupational carpal tunnel syndrome in Washington State, 1984–1988. *Am. J. Public Health* 81(6): 741–746.

Grandjean, E., 1997. *Fitting the Task to the Man*, Fourth Edition. New York: Taylor and Francis, Ltd.

Hagberg, M., and Wegman, D.H., 1987. Prevalence rates and odds ratios of shoulder-neck diseases in different occupational groups. *Br. J. Ind. Med.* Sept. 44(9):602–610.

Hoekstra, E.J., Hurrell, J.J., and Swanson, N.G., 1994. Hazard evaluation and technical assistance report: Social Security Administration Teleservice Centers, Boston, MA; Fort Lauderdale, FL. Cincinnati, OH: U.S. Department of Health and Human Services, Public Health Service, Centers for Disease Control and Prevention, National Institute for Occupational Safety and Health, NIOSH Report No. 92–0382–2450.

Kamon, E. and Goldfuss, A.J., 1978. In-plant evaluation of the muscle strength of workers, *Am. Ind. Hyg. Assoc. J.* 39:801–807.

Keegan, J.J., 1953. Alterations of the lumbar curve related to posture and seating. *J. Bone Joint Surg.* 35A(3) 592–601.

Kodak, 1983. Human Factors Section, Eastman Kodak Company. *Ergonomic Design for People at Work. Volumes I and II.* New York: John Wiley and Sons.

Leino, P., 1989. Symptoms of stress predict musculoskeletal disorders. *J. Epidem. Comm. Health* 43:293–300.

Lowry, C., Chadwick, R., and Waltman, E., 1976. Digital Vessel Trauma from Repetitive Impact in Baseball Pitchers. *J of Hand Surgery* 11:236-238.

Nachemson, A. and Morris, J.M., 1964. In vivo measurements of intradiscal pressure. *J. Bone Joint. Surg.* 46A: 1077.

National Electrical Contractors Association, 1989. *Overtime and Productivity in Electrical Construction.* Bethesda, MD: NECA.

NIOSH, 1994. *Applications Manual for the Revised Lifting Equation.* Department of Health and Human Services, Center for Disease Control, National Institute for Occupational Safety and Health. DHHS (NIOSH) Publication No. 94–110. Springfield, VA: U.S. Department of Commerce Technology Administration, National Technical Information Service.

Ohlsson. K., Attewell, R., Paisson, B., Karlsson, B., Balogh, I., and Johnsson, B. 1995. Repetitive industrial work and neck and upper limb disorders in females. *Am. J. Ind. Med.* 27(5):731–747.

Ohlsson, K., Attewell, R., and Skerfving, S., 1989. Self-reported symptoms in the neck and upper limbs of female assembly workers. *Scand. J. Work Environ. Health* 15(1):75–80.

Putz-Anderson, V., 1988. *Cumulative Trauma Disorders: A Manual for Musculoskeletal Disorders of the Upper Limb.* New York: Taylor and Francis.

Rempel, D., 1995. Musculoskeletal loading and carpal tunnel pressure. In: Gordon, S.L., Blair, S.J., and Fine, L.J., eds. *Repetitive Motions Disorders of the Upper Extremity.* Rosemont IL: American Academy of Orthopaedic Surgeons, pp. 123–133.

Rossignol., A.M., Morse, E.P., Summers, V.M., and Pagnotto, L.D., 1987. Video display terminal use and reported health symptoms among Massachusetts clerical workers. *J. Occ. Med.* 29(2):112–118.

Silverstein, B. A., 1985. *The Prevalence of Upper Extremity Disorders in Industry.* Ann Arbor: Center for Ergonomics, University of Michigan.

Stevens, J.C., Sun, M.D., Beard, C.M., O'Fallon, W.M., and Kurland, L.T., 1988. Carpal tunnel syndrome in Rochester, Minnesota, 1961 to 1980. *Neurology* 38:134–138.

Tichauer, E.R., 1978. *The Biomechanical Basis of Ergonomics.* New York: John Wiley and Sons.

Problem-Solving Process

MacLeod, D., 1999. *The Ergonomics Kit for General Industry.* New York: Lewis Publishers.

MacLeod, D., 1998. *The Office Ergonomics Kit.* New York: Lewis Publishers.

American National Standards Institute, ANSI Z-365 Committee, 1995. *Control of Work-Related Cumulative Trauma Disorders, Draft Appendix.* Itasca, IL: National Safety Council.

American National Standards Institute, 1988. ANSI/HFES Standard No. 100-1988. *American National Standard for Human Factors Engineering of Visual Display Workstations.* Santa Monica: Human Factors and Ergonomics Society.

Index